Greenway College

*How You Can Help Build the School
That Engineers Our Sustainable Future*

TROY MCBRIDE

Greenway College: How You Can Help Build the School That Engineers Our Sustainable Future

Published by Greenway Institute.

ISBN: 978-0692235591
LCCN: 2014910704

Contents

A Sustaining Vision

> Every revolutionary idea—in science, politics, art, or whatever—seems to evoke three stages of reaction. They may be summed up by the phrases: (1) "It's completely impossible—don't waste my time"; (2) "It's possible, but it's not worth doing"; (3) "I said it was a good idea all along."
>
> —Arthur C. Clarke, *The Promise of Space*, 1968

Greening Up

Somehow, certain technologies that once sounded suspiciously hippie-dippie—renewable energy, sustainability, and recycling—have become as mainstream as Old Glory. Despite a partisan divide on numerous issues, Americans today are largely united when it comes to renewable energy and other Earth-friendly technologies. For example, surveys by Pew Research Center,[1] USA Today/Gallup,[2] ORC International, the Public Policy Institute of California,[3] the National

---

1 Cary Funk and Meg Hefferon, "U.S. Public Views on Climate and Energy," November 25, 2019, accessed March 16, 2020, https://www.pewresearch.org/science/2019/11/25/u-s-public-views-on-climate-and-energy/.

2 Wendy Koch, "Polls: Americans Back EPA and Clean Energy," USA Today, February 7, 2011, accessed May 18, 2012, http://content.usatoday.com/communities/greenhouse/post/2011/02/americans-back-epa-solar-wind/1#.T1e8EJj59AY.

3 Mark Baldassare et al., "Californians and the Environment," Public Policy Institute of California, July 2011, accessed May 18, 2012, http://www.ppic.

Renewable Energy Laboratory,[4] and other groups all find *bipartisan* majorities in favor of wind and solar power. In a USA Today / Gallup poll, 83 percent of Americans said they wanted the government to put a high priority on incentives for renewable power—more than favored overhauling the federal tax code (76 percent) or withdrawing faster from Afghanistan (72 percent).[5] A Republican polling firm found that even recent partisanship had not weakened public support for clean energy: "Voters believe that the clean energy economy is here and is growing, and they want their state to have a part of it."[6] Decades of polls have shown that about 70 percent of Americans consistently support recycling,[7] and 83 percent of polled consumers said that a company's commitment to sustainability was "very or somewhat important" in their purchasing decisions.[8]

Did the hippies win our hearts and minds after all? They undeniably made their mark on matters from diet and music to education and religion, but it was the geeks—the engineers and scientists—who have for decades been steadily making

org/content/pubs/survey/S_711MBS.pdf.

4   "Consumer Attitudes About Renewable Energy: Trends and Regional Differences," NREL/SR-6A20-50988, Natural Marketing Institute for the National Renewable Energy Laboratory, April 2011, accessed May 18, 2012, http://apps3.eere.energy.gov/greenpower/pdfs/50988.pdf.

5   Koch, op. cit.

6   Dave Metz and Lori Weigel, "Recent Research Insights into the Solyndra Issue," Fairbank, Maslin, Maullin, Metz & Associates, September 26, 2011, accessed May 18, 2012, http://online.wsj.com/public/resources/documents/SolyndraPoll.pdf.

7   "Too Good to Throw Away: Recycling's Proven Record," Natural Resources Defense Council, accessed May 18, 2012, http://www.nrdc.org/cities/recycling/recyc/recyint.asp.

8   "Poll Shows Commitment to Sustainability Still Important to Consumers," Capstrat, accessed May 18, 2012, http://www.capstrat.com/insights/articles/poll-shows-commitment-to-sustainability-still-important-to-consumers/.

green technologies more affordable, effective, and common-place. In many sectors, green is the new normal.

Car vs. Cliff

Engineers apply the principles of physics to shape the machines and processes that make our industrial civilization possible. In its physical aspect, that civilization is itself a vast, elaborate machine that conjures the mechanical equivalent of ten laborers to serve each one of us. It has increased health and well-being for vast numbers of people over the past two centuries. Yet for all its power and ingenuity, and despite the recent rise of green technologies in many sectors, it is not "sustainable," and cannot run forever in its current form. Why not? We are consuming subsurface ore, oil, and gas at an increasing rate while at Earth's surface, there is less arable soil and clean water. The fossil fuels we take out of the ground we put into the air, creating pollution that kills millions of people every year[9] and is changing Earth's climate (as 97 percent of working climate scientists affirm).[10] Total human demands already exceed the long-term carrying capacity of the biosphere by 20 percent.[11] At this point, we bypass the hippies and ask the engineers, "Short of returning to a tribal subsistence lifestyle, how can we ensure that our descendants will live decently, or at all?"

That our high-resource society has generated so much health and wealth so far is encouraging, but does not prove

---

9   "Air Quality and Health," September 2011, World Health Organiza-tion, accessed May 24, 2012, http://www.who.int/mediacentre/factsheets/fs313/en/.
10   Justin Gillis, "Study Affirms Consensus on Climate Change," *New York Times*, June 22, 2010, accessed May 24, 2012, http://green.blogs.nytimes.com/2010/06/22/evidence-for-a-consensus-on-climate-change/.
11   Mathis Wackernagel et al., "Tracking the Ecological Overshoot of the Human Economy," *Proceedings of the National Academy of Sciences of the United States of America*, 99:14 (2002), 9266–9271, accessed May 24, 2012, http://www.pnas.org/content/99/14/9266.long.

that it can do so forever: a car accelerating toward a cliff edge performs superbly right up to the moment it launches into space (and perhaps a few seconds longer). How can our civilization preserve its gains without being doomed to crash and burn? We can figuratively hit the brakes—constraining consumption. But doing so could jeopardize some of the health and well-being achievements of the past two centuries. A better alternative would be to think like engineers and use applied science to make our civilization-machine *sustainable*. Build wings, bolt them on, and take off. Use technology to make the cliff irrelevant.

For too long, confronted with the evidence of trouble ahead, many people have shrugged and said, "Science will figure out something, someday," or "Good old American ingenuity will see us through." But those responses will no longer do, because "someday" has arrived. The good news is that science has figured out its "something": affordable renewable energy, sustainable energy storage and energy carriers, dramatic efficiency improvements, and elegant zero-waste practices. These are the technical and human arts of sustainability. These will be our wings.

Enter Greenway College

Young Americans are eager to be part of the solution: two-thirds of prospective college students surveyed for the *Princeton Review*'s "College Hopes and Worries" feature said that a college's "environmental commitment" would be a factor in where they applied.[12] According to *USA Today*, "College students are flocking to sustainability degrees, careers."[13]

---

12   "2011 College Hopes & Worries Survey Report," *The Princeton Review*, accessed March 13, 2012, http://www.princetonreview.com/uploadedFiles/Sitemap/Home_Page/Rankings/2011%20College%20Hopes%20and%20Worries%20Survey%20Report.pdf.

13   Jillian Berman, "College Students Are Flocking to Sustainability Degrees, Careers," *USA Today*, August 3, 2009, accessed February 10, 2012,

A four-year engineering and science college dedicated to this demand—an institution of learning devoted entirely to sustainable technology and engineering—would have unique appeal. As the first of its kind anywhere in the world, it would be a focus of excitement, passion, high hopes. It would attract research talent, teaching talent, the most promising students. It would be a thought leader, an idea generator, and living, working proof that a better world is possible through the power of applied knowledge.

To succeed, such a college would have to do more than shuffle modestly toward sustainability: a solar panel here, a LEED-certified building there, regular meetings of a sustainability committee, a ten-year plan to lower environmental impact. Such measures do not inspire passion. A sustainability college would have to be *amazingly* greener (while being at least as well lit, comfortable, productive, workable, and academically rigorous as any other school). It would have to not merely push the envelope but reinvent the envelope, then push it again, then reinvent it again, and again, and again. It would have to be a standing challenge to the imagination.

We propose to build such a school: Greenway College, a *zero-sacrifice, totally green* four-year college. Greenway will be a stand-alone learning community entirely devoted to expanding and disseminating the technological bases of a sustainable civilization and to graduating learners rigorously equipped to further this goal wherever they go. It will be a driver of the sustainability knowledge revolution.

And Greenway will not just talk the talk. It will apply on its own campus the knowledge base that it gathers and enriches, and will foster programs across the nation and planet to shift society toward sustainability. It will be a model, a seedbed, a driver, a transformation factory, a bottomless bag of profit-

---

http://www.usatoday.com/news/education/2009-08-02-sustainability-degrees_N.htm.

able tricks, and a gusher of insights into devices, materials, energy sources, and patterns of resource use and reuse.

People are going to love this place. With your help, we are going to build it. This book lays out how.

Four Goals, Three Tools

But first, a word about the "green" movement that has grown up since the 1960s and without which there would probably be no Greenway College. Despite our debt to the green subculture, the founders of Greenway College are not uncritical fans of it or any other way of thinking. In truth, there has been no monolithic movement toward greenness, no unified green ideology or cookbook. Rather, as Thomas Graedal has suggested, there is a loose, diverse cluster of movements, most of whose members share four basic goals: "1) maintaining the existence of the human species, 2) maintaining the capacity for sustainable development, 3) maintaining the diversity of living things, and 4) maintaining the aesthetic richness of the planet."[14]

Not many people would reject any of the Four Goals, though people can and do argue endlessly about how to achieve them. No wonder such ideas have broad bipartisan support, even though 60 percent of Americans don't want to be called "environmentalists."[15]

The Four Goals have generally been pursued using three main tools, namely environmental legislation, simpler living, and technological innovation. The first is clearly indispensable, though the proper form and degree of legislative intervention can be debated: without some regulation, any polluter would be free to externalize their costs to whoever happened to be downstream and downwind, "costs" not just in dollars

---

14   Thomas Graedal, *Streamlined Life Cycle Assessment* (New Jersey: Prentice Hall, 1998).
15   Gillis, op. cit.

but in disease and death. A good law saves lives. Each year, reductions in air pollution from the 1990 amendments to the Clean Air Act—just one aspect of just one successful environmental law—prevent over 130 thousand heart attacks, 1.7 million asthma attacks, and 13 million lost workdays. The economic benefits of major environmental laws have wildly outweighed their costs, thanks primarily to savings in medical expenses.[16]

Yet legislation is inherently political: parties contend, regulations trigger pushback campaigns, enemies are made. The discourse around environmental regulation has become astonishingly bitter in the United States today. Greenway must not contribute to this bitterness or become a victim of it. Though its staff and students will, we hope, be active in democratic politics across a wide spectrum of views, the college itself must remain truly neutral. It must authentically welcome persons with any and all beliefs that are compatible with the basic standards of liberty and responsibility prevalent in our society. It will not be a liberal or conservative clubhouse, but constitutionally nonideological and nonpolitical.

The next popular tool for achieving the Four Goals is simpler living. We cannot utterly evade the question of lifestyle, though some people would prefer to. Is sustainability compatible with everyone on the planet having a house, an acre, and a car? How about three houses, two hundred acres, three cars, and a yacht? How much is too little, and how much is too much, and who gets to decide? One person's liberating simplicity is another's bitter sacrifice. What to one person is a mere technical tweak is to another an intolerable attack on personal liberty: consider the varied reactions to the transition from incandescent to compact fluorescent to LED lightbulbs.

---

16   US Environmental Protection Agency Office of Air and Radiation, "The Benefits and Costs of the Clean Air Act from 1990 to 2020," March 2011, accessed March 6, 2012, http://www.epa.gov/air/sect812/feb11/fullreport.pdf.

Building a dormitory or a dining hall is, then, at some level an unavoidably ideological act, because choices must be made and every choice implies some philosophy of what is decent or enough, and somebody, somewhere, is bound to disagree. By taking any specific form, by merely existing, Greenway is bound to please some and offend others.

Nevertheless, within realistic limits, Greenway College shall avoid lifestyle advocacy. We have no quarrel with persons who wish to consume less and persuade others to do the same, but our mission is not lifestyle change any more than it is legislative activism. Greenway is dedicated to the third tool, the technological approach, which is based on the assumption that there is, globally speaking, enough to go around. We maintain that within certain broad limits, and if we make intelligent use of its resources, this planet is capable of providing light, warmth, education, comfortable roomy spaces, hot showers, tasty food and drink, vivid entertainment, clean air and water, and vast wilderness to all its people. This is what we mean by "zero sacrifice."

As an institution, therefore, Greenway will advocate not lifestyle change but *technical responses to technical opportunities.* Society wishes certain tasks to be done, certain services to be provided; very well; what is the absolute best way to fulfill those wishes? How can we supply an affordable, comfortable, industrial way of life with zero waste, sustainably, indefinitely, and in a manner that does not leave our grandchildren standing empty-handed atop a destroyed planet?

This book presents our vision and invites you to join us in making that vision a reality.

Glancing Ahead

In chapters 2, 3, and 4, we review relevant green technologies for energy generation, energy storage, and zero-waste engineering. We describe how these can be incorporated into the fabric of Greenway College. In chapter 5, we lay out our vision of a real school, fleshing out a proposal for a totally green stand-alone campus that can be readily achieved with technologies available today (yet remain capable of endless improvement). Chapter 6 reviews college operations: how will Greenway be governed; what sort of school will it be? Chapter 7, last but not least, talks about money.

We hope that by the end of this book you will be as excited as we are about Greenway and as convinced that its determination to be zero sacrifice, zero waste, and self-sufficient in energy is anything but pie in the sky. We also hope that you will choose to become a part of this historic effort. There has never yet been a college like Greenway, but the time is ripe. Soon, with your help, Greenway will be spreading fresh, market-ready technologies, principles, and graduates across our nation and planet—helping to engineer our sustainable future.

Making and Using Energy

It is in our vital interest to diversify America's energy supply—and the way forward is through technology.
—President George W. Bush, State of the Union address, January 23, 2007

Energy will be the immediate test of our ability to unite this nation, and it can also be the standard around which we rally. On the battlefield of energy we can win for our nation a new confidence, and we can seize control again of our common destiny.
—President Jimmy Carter, July 15, 1979

We are like tenant farmers chopping down the fence around our house for fuel when we should be using Nature's inexhaustible sources of energy—sun, wind and tide. . . . I'd put my money on the sun and solar energy. What a source of power! I hope we don't have to wait until oil and coal run out before we tackle that.
—Thomas A. Edison, 1931

The Universal Ingredient

Without energy nothing is harvested, mined, refined, manufactured, packaged, or shipped. Energy cools and lights our buildings, processes our data, powers our tools, lights our streets, propels our vehicles, and runs our farms and factories. Consequently, when energy prices go up, the price of everything goes up, from toothpaste to trans-Pacific container shipping. Even modestly higher-than-normal prices for just one source of energy, crude oil, can be a drag on the entire economy.[1]

All the improvements in physical well-being achieved by industrialized society over the last 150 years—like greater longevity, leisure, and comfort—have been made possible by abundant, concentrated energy supplies. All our food, housing, transport, communication, and military might is built and animated by energy. At first that energy came primarily from wood, then coal, then coal and oil, and now coal, oil, natural gas, and a potpourri of nuclear, hydro, wind, solar, and other sources. The oil embargo of 1973–74 highlighted how tightly national security depends on energy security, while rubbing our noses in the fact that our "addiction" to oil, as President George W. Bush called it,[2] leads inexorably to energy *in*security. The Department of Defense says that energy sustainability is an "organizing paradigm that applies to all DoD mission and program areas" and that the US military's "heavy reliance on fossil fuels creates significant risks and costs at a tactical, as well as a strategic level."[3] The main remedy, the DoD says, "will be to reduce reliance on fossil fuels through energy efficiency and renewable energy." Nevertheless, the US military is still the world's largest single user of petroleum.[4] It is easier to see the need to change than to change.

---

1   R. E. Earley and K. Smith, "What Has Happened to the Share of Energy in the U.S. Economy Since the Early 1970s?" US Energy Information Agency, updated April 10, 2001, accessed May 21, 2012, http://www.eia.doe.gov/oiaf/economy/energy_price.html.

2   "Bush: U.S. Must Break Oil 'Addiction,'" *CNN*, February 11, 2009, accessed May 21, 2012 http://www.cbsnews.com/2100-250_162-1260701.html.

3   *Strategic Sustainability Performance Plan* (FY 2010), US Department of Defense Office of Installations & Environment, accessed May 21, 2012, http://www.denix.osd.mil/sustainability/upload/DoD-SSPP-PUBLIC-26Aug10.pdf.

4   S. Karbuz, "Can the US Military Move to Renewable Fuels?" *Bulletin of the Atomic Scientists*, October 31, 2008, accessed May 21, 2012, http://

The litany of environmental harms caused by extract-
ing and applying energy is all too familiar. Obtaining fossil
fuels entails destructive mining practices (e.g., mountaintop
removal), drilling both onshore and off, hydraulic fracturing,
and strip mining (fully exploiting Alberta's tar sands would
entail disturbing a wilderness area bigger than Florida).[5]
Nuclear power suffers from unique waste-disposal problems
and is tied to nuclear weapons proliferation because refined
uranium and plutonium are the stuff of ultimate weapons.

Over a trillion dollars is spent on energy per year in the
US.[6] Despite that awesome sum, we are in the era of cheap,
abundant energy and have been since the dawn of the industrial revolution. Experts debate how long the cheap fossil
fuel portion of this party will last, but most agree that the
current system is unsustainable: we cannot go on like this
forever. At projected worldwide resource consumption rates,
the Earth's known conventional fuel reserves amount roughly
to about forty years of oil, fifty-four years of natural gas, over
one hundred years of coal, and over two hundred years of
uranium 235 (at present consumption rates, not proposed
nuclear-renaissance rates).[7] The exact dates are debatable,

www.thebulletin.org/web-edition/features/can-the-us-military-move-to-re-
newable-fuels.

5   "Strip Mining for Oil in Endangered Forests," Natural Resources Defense
Council, June 2006, accessed May 21, 2012, http://www.nrdc.org/media/
docs/060607a.pdf. See also "Alberta Tar Sands: It's Not Just About Mining,"
Skytruth, accessed May 21, 2012, http://blog.skytruth.org/2011/11/and-t-
only-gets-larger-from-here-on-out.html.

6   "Consumer Expenditure Estimates for Energy by Source, 1970–2009,"
US Energy Information Administration, accessed March 16, 2012, http://
www.eia.gov/totalenergy/data/annual/pdf/sec3_11.pdf.

7   "2010 Survey of Energy Resources," World Energy Council, accessed
May 21, 2012, http://www.worldenergy.org/documents/ser_2010_report_1.
pdf. Also: Steve Fetter, "How Long Will the World's Uranium Supplies Last?"
Scientific American, accessed May 21, 2012, http://www.scientificamerican.
com/article.cfm?id=how-long-will-global-uranium-deposits-last.

but do not really matter: some are less than a generation away and none are extremely remote.

Alternative or "renewable" energy generation, in contrast, relies on energy flows powered by the sun and the Earth's inner heat, all of which are inexhaustible on any timescale relevant to human history. Sunlight is the main source of renewable energy, whether directly (in heat collection or electricity generation), the winds (via differential heating of the Earth's surface), the water cycle (hydroelectricity), or plant life (wood and other biomass). The sun showers the Earth with as much energy every hour as human civilization uses every year.[8] The motions of the Earth, sun, and moon drive ocean tides, which can also be harnessed for power generation, while heat left over from the formation of the solar system, supplemented by the slow breakdown of uranium 238 in the Earth's deep interior, drives geothermal power. Renewable energy sources are therefore "sustainable": they last. As a major side benefit, they tend to free their users from far-flung supply lines, divorcing energy supply from geopolitics.

Most energy experts agree with the majority of the American people: we need to abandon today's dirty, fragile, expensive, and ultimately doomed energy system in favor of a clean, secure, affordable, sustainable one. To move toward this goal we need to both use energy as efficiently as possible and produce what we do use in a sustainable manner. Greenway College will walk both paths at once. Without sacrificing any of the comforts and conveniences of a modern college campus, it will be both a *super-efficient user* and a *self-sufficient supplier* of energy.

---

8   Oliver Morton, "Solar Energy: A New Day Dawning?" *Nature* 443 (2006), 19–22, accessed May 24, 2012, http://www.nature.com/nature/journal/v443/n7107/pdf/443019a.pdf.

Money vs. Mission?

In terms of up-front cost, simply hooking up to the electrical grid in the usual way would be cheaper than attaining energy self-sufficiency. The latter requires efficient energy use coupled with on-site generation and energy storage. But a college is not a factory: it does not exist to make a given amount of profit per widget sold or student graduated. Although a college must survive financially, it has a goal, a reason for existing, that transcends its budget: namely, to foster learning. The unique educational mission of Greenway College is to *produce graduates that will engineer our sustainable future and contribute to our graceful transformation to a sustainable society*. It will equip its students to be first-rate users, innovators, and entrepreneurs of the whole range of sustainable technologies by involving them in the exploration, refinement, and demonstration of exactly those technologies in the fabric of the campus itself.

Greenway's energy system will, then, be not only a supplier of energy services but an on-campus teaching resource, and must be planned, budgeted, and evaluated accordingly. On these terms, it does not make sense to ask which energy technologies will cost the least merely to install, or which technologies will pay for themselves in a given time window, and then build Greenway around the answers to such questions. The question for Greenway is, *what affordable mix of energy technologies will most fully and flexibly serve the college's reasons for existing?* This mix will likely include diverse renewable energy sources, storage technologies, and a wide array of efficiency and sustainability solutions.

What's more, room for change will be built into Greenway's energy system. The college's mission implies that it must reinvent itself over time, as students and faculty learn, experiment, and build. It may begin, for example, by relying heavily on photovoltaics and biomass, then add wind, small-scale hydropower, geothermal energy, or other technologies as time goes on. It may begin by being self-sufficient in energy, then become capable, over a decade or two, of exporting energy to surrounding communities or the grid.

Energy wisdom boils down to using energy efficiently and making it well. Below, we outline how Greenway can economically do both.

Using Less Without Doing Less

Energy efficiency is based on the traditional values of thrift and common sense. No sensible person opens all their windows in midwinter and burns extra fuel to heat the great outdoors. If one can achieve equal or greater comfort with less energy at lower cost, one always chooses the more efficient path.

Energy efficiency reduces energy demand. Demand can also be reduced by changing lifestyles (taking shorter showers, say), but most people would rather keep getting all the energy services they are accustomed to, only more efficiently (by getting their domestic hot water from a heat pump, for example). As explained in chapter 1, we tend to agree: efficiency is a less controversial path than lifestyle change. It appeals to enlightened self-interest rather than asking people to make what they perceive as sacrifices. Therefore, Greenway will not be austere. If anything, it will be a paragon of normality and comfort. This can be done because most people care not about kilowatts and kilowatt-hours, but energy *services*. They don't want amps, volts, watts, and joules but comfortable rooms, pleasant lighting, hot coffee, cold milk, and computers that work. If these demands can be met affordably with less energy—and they can—it makes every kind of sense to do so.

We cannot overemphasize that Greenway will provide a normal array of comforts and energy services. Our mantra is "totally green with *zero sacrifice*."

It might be objected, with some justification, that "sacrifice" is subjective: to live like a millionaire might involve "sacrifice" for a billionaire. But we use "zero sacrifice" to mean simply a standard of living that the vast majority of Americans would recognize as comfortable. We will assure a stable, reliable energy supply sufficient for all the activities of any college campus—reading, lab work, homework, computing, eating, socializing, showering, and so on. One or two scions of privilege might grumble, but we think it reasonable to describe such a campus as "zero sacrifice."

Totally green zero sacrifice requires high efficiency. The opportunity is huge: our society's usual methods of using energy are grossly wasteful, bleeding billions of dollars per year from homes and corporations. A report by McKinsey & Company found that technological adjustments alone—no lifestyle change—could cut US non-transportation energy use by 23 percent while saving half a trillion dollars, net.[9] And this is a *conservative* estimate, based on minor adjustments like using more efficient light bulbs, turning off unused appliances, weather-stripping windows and doors, and so forth. A campus designed from the ground up to provide lighting, heating, cooling, computing, and other energy services with the least possible energy might cost-effectively cut energy use by half, or even more. With demand sufficiently low, and with careful and deliberate planning, such a campus can readily produce all of its own energy without relying on the grid or deliveries of fuel.

Making Energy

It is, using traditional practices, much simpler to build a conventionally powered building than one that produces its

---

9   H. C. Granade et al., "Unlocking Energy Efficiency in the U.S. Economy: McKinsey Global Energy and Materials, 2009," accessed May 21, 2012, http://www.mckinsey.com/Client_Service/Electric_Power_and_Natural_Gas/Latest_thinking/Unlocking_energy_efficiency_in_the_US_economy.

own energy. In the usual course of affairs, one subcontractor digs and pours a standard foundation, another frames the building to comply (minimally) with code, another trucks in out-of-the-box coolers, heaters, windows, insulation, circulators, and other components, and another installs gas lines or oil tanks. An electrician runs the on-site wiring and utility workers hook it up to the grid. Finally, like wedding guests at the end of the reception, they all walk away and leave you with your new life.

Simple, maybe, but expensive, despite the fact that everything is done by lowest bidders. None of these contractors are motivated to reduce your energy costs, because they aren't going to be paying your electric, gas, and oil bills for all the years that your new building will need constant infusions of energy to remain livable. You are roped to those bills—bills that are tied to the vagaries of OPEC policy, peak oil, carbon pricing, regulation, Middle East crises, Enron-style debacles, and a dozen other factors out of your control. Short of major retrofits to your building and its gear, you *must* pay those bills, no matter how high energy prices go, because without a nonstop flow of ample energy from outside sources, your asset is a cold, dark shell.

At Greenway we will make our own energy, completely separating ourselves from the mainly fossil-fuel-powered grid and transportation network, from fuel-price fluctuations, and from a host of related environmental and geopolitical liabilities. Among other benefits, this will buffer Greenway against the vicissitudes of the larger economy, making us fiscally nimbler, better equipped to thrive in hard times, slower to sink, quicker to rise.

What would you do if you could start with a blank sheet of paper and design your own campus from scratch, using all the best technologies available today? Even if you are not an expert, your approach will probably improve greatly over today's out-of-the-box construction practices. We invite you to put your bookmark in for a few minutes and sketch out the elements of a design, and then consider our approach, outlined below. You will probably think of many of the same basic principles that we intend to apply.

Most basically, "energy" is not just one thing: different forms of energy tend to fulfill different needs, heating and cooling buildings, powering lighting and electrical devices, and moving machinery and vehicles.

Buildings. About 40 percent of present US energy use occurs in buildings—heating and cooling them and powering all the activities that go on inside.[10] When it comes to cutting this use, radical is the only realistic. US energy secretary and Nobel Prize winner Stephen Chu has said that for new construction, "extremely cost-effective buildings with energy savings of 60–80 percent are possible" without sacrifice of comfort.[11] The US National Renewable Energy Laboratory recently built three large commercial-class office spaces in Colorado that generate as much energy as they use—"zero net energy" buildings.[12] Inside such structures, a wide array of technically mature, field-tested, affordable tricks trims energy demand without sacrificing services: daylighting, LED lighting, timers, motion detectors, heat pumps, and much more. For heating and cooling of new buildings, the job can often be done for a tenth or less of current energy use, according to the InterAcademy Council—not speculatively, but as already demonstrated in numerous structures.[13] Over twenty-five thousand build-

---

10   "Buildings and Their Impact on the Environment: A Statistical Summary," US Environmental Protection Agency, April 22, 2009, accessed March 16, 2012, http://www.epa.gov/greenbuilding/pubs/gbstats.pdf.

11   "Secretary Chu Op-Ed on Energy Efficiency from the World Economic Forum," March 16, 2010, accessed March 16, 2012, http://energy.gov/articles/secretary-chu-op-ed-energy-efficiency-world-economic-forum.

12   "Research Support Facility: Leadership in Energy Performance," NREL, accessed March 16, 2012, http://www.nrel.gov/sustainable_nrel/pdfs/51742.pdf.

13   "Lighting the Way: Toward a Sustainable Energy Future," InterAcademy

ings requiring almost no heating or cooling, even in northern-European and Mediterranean climates, have been built to the European "Passivhaus" standard.[14]

But if such big savings are so easily realized, why are only a select few companies and individuals reaping the benefits of aggressive efficiency programs? The causes of this strange, almost anticompetitive, wastefulness include irrational regulations, institutional inertia, split incentives, and technical ignorance. The efficiency resource is intrinsically fragmented, meaning that to exploit it requires thoughtful action at many points: it is *simpler* to pay a few fat energy bills than to use less energy. (But then, it is always simpler to lose money than to earn it.) Another chronic stumbling block is that efficiency investments pay back only over time, not in one bright flash by throwing a giant green switch at a ceremony. Also, we have had a tendency in recent decades to focus on short-term financial motivators, but efficiency takes time to pay.

Electricity. Electricity is the most broadly useful form of energy, though expensive to produce and store. Average US electric power usage per adult is about 30 kilowatt-hours (kWh) per day—enough to keep about fifty laptop computers or two hair dryers running nonstop.[15]

Most of us don't generate our own electricity but buy it from the national grid, which is that network of power plants,

---

Council, October 2007, accessed May 21, 2012, http://www.interacademy-council.net/File.aspx?id=24548.

14   Tom Zeller Jr., "Can We Build in a Brighter Shade of Green?" *New York Times*, September 25, 2010, accessed May 21, 2012, http://www.nytimes.com/2010/09/26/business/energy-environment/26smart.html?pagewanted=all.

15   A kilowatt (kW) is a unit of energy *rate* or *flow*, similar to gallons per minute for a liquid. A kilowatt-hour (kWh) is a unit of energy *quantity*, like gallons. A single thousand-watt hair dryer uses 1 kW of power; left on for one hour, it uses 1 kWh of energy. As of 2008, the average US residential customer used 11,040 kWh of electricity per year (http://tonto.eia.doe.gov/ask/electricity_faqs.asp).

Energy, Power, and Units

Writers often use the words "energy," "power," and "electricity" interchangeably. This causes much confusion in discussions of energy policy, efficiency, and related topics.

In science, **energy** is the ability to do a certain amount of work. A charged battery, for example, contains a fixed amount of potential energy. (It remains "potential" until it actually does something.)

Power is the rate at which energy is used to perform some task, like lighting a light bulb. Consider using a battery to run a light bulb. A brighter light bulb uses up the energy in the battery faster than a dimmer bulb, turning it into light and heat, so we say that it uses more power.

Electricity is a vague word, therefore not often used by engineers or scientists, but in everyday speech it refers simply to that invisible, amazingly useful stuff that comes from wall sockets. A decent nontechnical definition of electricity, in this everyday sense, would be **electrons doing useful work**. One can speak of both electrical **energy** (some amount of energy made available in the form of electricity, typically measured in kilowatt-hours) and electrical **power** (the rate at which electrical energy is used, typically measured in kilowatts).

Electrical energy is only one form of energy. One can also have chemical potential energy (e.g., in coal or gasoline), thermal energy (e.g., for heating), kinetic energy (e.g., the energy of running water), and others. Different forms of energy tend to be used for different tasks—gasoline in cars, or electrical energy for computers, say.

Electricity is unique because it can be used to accomplish almost any task. It can heat soup or power a laser beam, cool a house or run a car. Electrical energy can even be stored (in effect) by charging a battery, compressing air, making hydrogen from water, or other methods.

Electricity is the direct, natural product of solar panels and wind turbines. At Greenway, we will make full use of electricity's versatility to provide energy services and expand our technology options.

long-distance power lines, local poles and lines, and other gear that produces and distributes electricity across the continent. Most electricity delivered through the grid comes from heavy-duty power plants: coal (around 31 percent), natural gas (around 29 percent), nuclear (around 22 percent), and hydroelectric (around 7 percent).[16] We shouldn't take our electricity for granted just because these large power plants are typically out of sight and out of mind. When the author was a professor, he led tours to visit many power plants: they are both engineering marvels and, with their immense hunger for resources and wastefulness of primary energy, clear examples of why sustainable technology and engineering are so important.

It is increasingly common (though still the source of a small percentage of US electricity) for customers to own grid-connected generators of their own such as solar panels, often set up to feed power to the grid whenever their output is not needed by the customer. Connecting one's home, business, or campus to both the grid and a small-scale generator is straightforward in the forty-seven states that now require power companies to accept interconnection with such installations and to buy surplus power from them.[17]

Grid connection has two major advantages for small-scale users and generators of power: First, any surplus on-site generation earns income. Second, if one's on-site power source fails, the grid can supply one's electricity needs. At the same time, on-site generation can keep one's lights on during a grid outage (though this is not standard practice for grid-connected systems). Grid outages, most of them lasting just

---

16   U.S. Energy Information Administration, "Electric Power Sector Energy Consumption," *Monthly Energy Review*, July 2019, accessed August 17, 2019, https://www.eia.gov/totalenergy/data/monthly/pdf/sec2_13.pdf.
17   "State Net Metering Policies," National Conference of State Legislators, accessed Aug 7, 2019, http://www.ncsl.org/research/energy/net-metering-policy-overview-and-state-legislative-updates.aspx.

long enough to crash every computer in the building, cost
US commercial-class electricity customers about $57 billion
every year.[18]

But if one is connected to the grid, one is implicated in
how the grid is powered, and that is a problem. As mentioned,
over 30 percent of US electricity is coal-generated (less in
some regions, more in others). The rest comes mostly from
natural gas, nuclear power, hydroelectricity, and wind (the
latter still only about 7 percent as of 2018, though growing).
If Greenway College wants to lead the way to a sustainable
clean-energy future, it should *create* that future by discon-
necting from the polluting grid while still providing itself
with a full range of energy services. To maximize its impact,
Greenway must walk the walk as well as talk the talk. When
people see a modern campus powered entirely and reliably by
renewable, locally generated energy, they will understand that
renewable energy can work on a national scale.

In the past, off-grid systems have tended to be underpow-
ered, one-of-a-kind designs, but when a stand-alone community like Greenway College is built from scratch with modern
know-how, its energy system can be designed to supply power
that is just as ample, reliable, and high quality as power from
the grid. Connection to the grid might be considered as a
way to sell surplus energy and for research purposes in the
future—but not as a backup or lifeline. That would be too
timid, too par for the course. It would not prove that renewables can stand on their own.

The arguments for being able to operate separately from
the grid seem to us compelling. We therefore propose to
produce all of Greenway's power on-site using a diverse portfolio of clean energy sources whose joint reliability will be

---

18   K. H. LaCommare and J. H. Eto, "Understanding the Cost of Power Interruptions to U.S. Electricity Consumers," Ernest Orlando Lawrence Berkeley National Laboratory, September 2004, accessed May 21, 2012, http://
certs.lbl.gov/pdf/55718.pdf.

higher than that of any one component. What distinguishes such sources is that they are *local and sustainable*, that is, harvest power from energy flows occurring naturally around and within the infrastructure of the community. Later in this chapter, we lay out exactly what Greenway's portfolio of clean energy sources will look like, including numbers on size and cost.

At this point you may be tempted to express skepticism. "Sure, 100 percent renewable electricity generation is possible when the wind is blowing and the sun is shining, but what about the rest of the time? What happens on a calm night?"

Good question. Here are quick, partial answers:

1. *Complementary sources.* Not all renewable power generators are intermittent. In addition to wind and solar, one may include around-the-clock generators in the power mix, such as a small hydropower turbine or a biogas-fired generator.

2. *Demand response.* During dips in power supply, one can arrange corresponding dips in demand by postponing power delivery to large loads that can wait a little while (e.g., freezers). At large scale, this technique is increasingly applied by US utilities across the grid.[19] Greenway College's intelligent, computerized power-management system will match select loads tolerant to postponement with power availability in a manner that is invisible to residents.

3. *Storage.* Energy can be put into storage when it is abundant and retrieved when production sags. This is already done by utilities, who most often use high-elevation hydroelectric water reservoirs for the

---

19   "2010 Long-Term Reliability Assessment," North American Electric Reliability Corporation, 2010, accessed May 21, 2012, http://www.nerc.com/files/2010_LTRA_v2-.pdf.

purpose. Smaller-scale storage may use batteries, reservoirs, compressed air, flywheels, synthetic liquid and gas fuels (e.g., hydrogen), and other means. With adequate storage, an off-grid installation powered entirely by erratic sources can have steady, around-the-clock power.

Greenway College will have to store energy in various forms to run on-campus vehicles, lawnmowers, portable devices, and the like. Storage will smooth the flow of renewable energy to these applications and enable Greenway to be off the grid, killing two birds with one stone. Storage is so important, in fact, that we devote the next chapter of this book to it.

4. *Transport.* Most vehicles run on gasoline and petroleum-derived diesel fuel, but one can readily purchase electric, methane-powered, hybrid biodiesel, and hydrogen vehicles and equipment for transportation and (with some relatively simple modifications) maintenance. Greenway's vehicular fleet will contain a mix of such technologies and evolve over time as such technologies evolve.

- Electric vehicles are an obvious option because Greenway will be producing its own electricity. These are efficient users of energy, with battery-based versions transferring about 50 to 60 percent of the electric energy used to the job of moving the car itself.[20] (This compares favorably with gasoline and diesel vehicles, which have tank-to-wheel efficiencies of around 20 percent.[21])

---

20   Battery storage and retrieval is around 80 percent efficient (90 percent charging, 90 percent discharging), the electric motor is about 85 percent efficient, and drivetrain efficiency is about 75 percent. This gives about a 50 percent overall efficiency. See http://www.electroauto.com/info/pollmyth.shtml.
21   "Electric Vehicles," US Department of Energy, 2012, accessed May 21,

- Methane burns cleanly, and methane-powered buses, forklifts, and other vehicles are commonplace today, so this is not an exotic technology either. Methane is produced naturally by anaerobic bacteria digesting organic matter, so methane from on-campus biodigesters will likely be available for vehicular and other purposes (e.g., cooking). Methane may also be synthesized by various means. Similar to the production of methane from biodigestion, biodiesel to fuel hybrid-diesel vehicles may be fermented from organic sources such as soy, canola seed, and algae.

- Hydrogen can be easily synthesized using electricity to separate the hydrogen and oxygen atoms in ordinary water (H_2O). Hydrogen can power vehicles either in fuel cells or in internal combustion engines. Hydrogen-powered vehicles, like biodiesel and biomethane vehicles, offer the convenience of rapid fill-up (seconds or minutes, versus hours to fully charge an electric car). However, if one's original energy form is electricity (e.g., from solar panels or wind turbines), hydrogen-powered vehicles are less efficient energy users than electric vehicles, having only 20 percent overall efficiency from outlet to wheel.[22] This is because hydrogen must first be *manufactured* using electricity, then reacted in a fuel cell or burned in an engine, with energy losses at every step, whereas an electric vehicle can

---

2012, http://www.fueleconomy.gov/feg/evtech.shtml.

22   Joe Romm, "Climate and Hydrogen Car Advocate Gets Almost Everything Wrong About Plug-in Cars," October 6, 2009, accessed May 21, 2012, http://thinkprogress.org/climate/2009/10/06/204758/climate-and-hydrogen-car-advocate-gets-almost-everything-wrong-about-plug-in-cars/.

charge from the electric source directly. Fueling a Greenway College on-campus fleet of twenty electric vehicles would require average electric generation of about 480 kWh/day; for hydrogen vehicles, about twice as much.

Costs

Alternative-energy installations often have higher up-front capital costs than conventional fossil fuel plants of equivalent power output, but also have the world-changing advantage of zero fuel costs (or, for biomass, fuel costs remain but are localized). Once all costs are factored in, the price per unit of energy produced by alternative generation, even with today's technology, becomes competitive.

As shown in table 2.1, wind installations are similar in price (per unit energy produced) to new gas turbines, and less expensive than a new coal plant, primarily because of wind's zero fuel costs. Biomass technologies can compete with conventional generation, depending on biomass fuel costs; biomass fuel costs can be near zero for on-site waste-mass usage or may include substantial farming costs.

The costs of solar electric installations have been declining exponentially for many years, and are likely to continue doing so.[23] New solar installations can be not only significantly less expensive than coal plants but competitive with new natural gas and wind installations.[24] Once installed, solar photovoltaic systems have minimal operational costs, making them an excellent investment. (Solar hot water units,

23   Dylan McGrath, "Study: U.S. Photovoltaic Costs Declining," *EE Times*, February 19, 2009, accessed May 21, 2012, http://www.eetimes.com/electronics-news/4081427/Study-U-S-photovoltaic-costs-declining.
24   "Levelized Cost and Levelized Avoided Cost of New Generation Resources in the Annual Energy Outlook 2020," U.S. Energy Information Administration, accessed June 11, 2020, https://www.eia.gov/outlooks/aeo/pdf/electricity_generation.pdf.

Table 2.1
Approximate cost data for new
energy generation installations

Technology	Levelized Capital Cost (¢/kWh)	Operational Costs including Fuel (¢/kWh)	Capacity Factor	Total Levelized Cost of Energy (¢/kWh)
Natural Gas	0.8	3.0	87%	3.8
Coal	4.8	2.9	85%	7.6
Nuclear	5.6	2.6	90%	8.2
Hydroelectric	3.7	1.6	75%	5.3
Geothermal	2.0	1.7	90%	3.8
Biomass	4.0	5.6	83%	9.5
Solar CSP	11.8	1.2	57%	13.0
Solar Photovoltaic	2.6	1.0	29%	3.6
Wind	3.0	1.0	41%	4.0

Source: U.S. Energy Information Administration, "Levelized Cost and Levelized Avoided Cost of New Generation Resources in the Annual Energy Outlook 2020."

as opposed to photovoltaic—direct solar to electricity-generating solar panels—are also competitive for their purpose.[25]) Both solar and wind costs per unit energy for new installations have dropped steeply over the last thirty years, as shown in figure 2.1.

What about nuclear power? We take no ideological stance on nuclear power, but new nuclear construction is currently more expensive than the major renewables. As shown in table 2.1, the DOE EIA 2020 reports a projected new-construction nuclear cost of 8.2¢/kWh as compared to 4.0 ¢/kWh and under for new wind and solar construction.

---

25   "Solar Hot Water Resources and Technologies," US Department of Energy, 2012, accessed May 21, 2012, http://www1.eere.energy.gov/femp/technologies/renewable_shw.html.

For Greenway, nuclear power is not an on-site option in any case: tiny nuclear reactors have been proposed, but are not commercially available.

Historical Levelized Cost of Energy

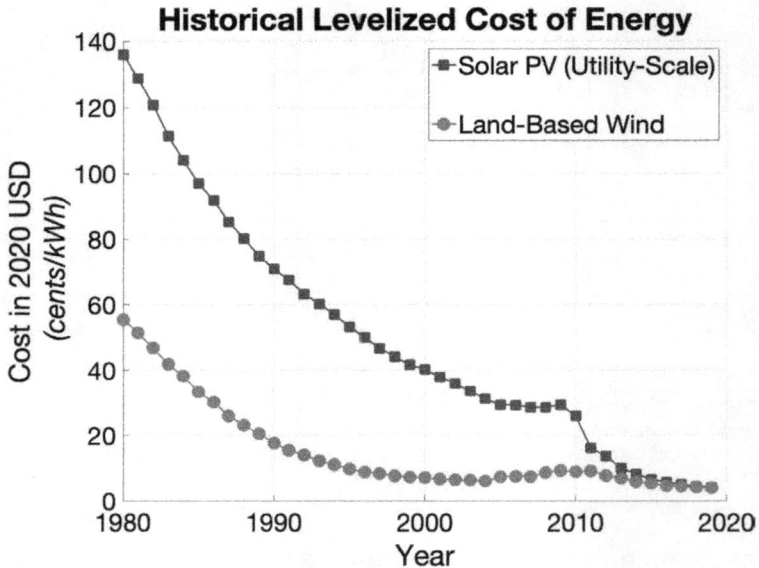

Fig. 2.1. Over the last thirty years, costs per kilowatt-hour for solar and wind power have fallen drastically.[26]

Making Energy at Greenway

As the previous section shows, cost is not an obstacle. The only question is exactly how to design a right-sized, afford-

---

26   Data to 2005 is from Melissa A. Schilling and Melissa Esmundo, "Technology S-curves in Renewable Energy Alternatives: Analysis and Implications for Industry and Government," *Energy Policy* 37:5(2009), 1767–1781, accessed May 21, 2012, doi.org/10.1016/j.enpol.2009.01.004.

Wind and solar data after 2005 are from *Lazard's Levelized Cost of Energy Analysis—Version 13.0*, November 2019, https://www.lazard.com/media/451086/lazards-levelized-cost-of-energy-version-130-vf.pdf.

able system that meets Greenway's needs—but that is a big question. In this section, we outline such a system.

With zero sacrifice, we conservatively project a 75 percent reduction in heating and cooling energy for buildings (compared to an otherwise comparable, conventionally powered college) as well as 50 percent reduction in all other forms of energy use. Eventually, additional savings may be realized, but this is a technically reasonable starting point.

For a residential college, typical energy requirements are around 25 to 50 kWh/day per person (lumping all power types together—electricity and fuels—and using the standardized kilowatt-hour unit, though this is often reserved for electricity). With efficiency improvements, energy requirements may be initially predicted at 10 to 25 kWh/day per person plus a margin for transportation and maintenance equipment. For a two hundred-person college, we conservatively estimate an around-the calendar average of about 4,000 kWh/day.

However, that is a lowball estimate. We require a system with ample supply (including storage—see chapter 3) to ride out periods of low generation and high demand. Electricity surpluses will not be a problem, but electricity shortages would most definitely be a problem. We therefore propose to build a 15,000 kWh/day electrical system, over three times the minimum electric generating capacity for our target campus population, that will provide 100 percent reliability, cover energy-storage losses, and accommodate future growth of the college. In routine campus operations, surplus power from generators will be used to build up long-term storage (e.g., hydroelectric, battery, or hydrogen). The choice of technologies for this generously-sized system will be matched to the geographic and topographic location of the college, although solar photovoltaic will almost surely be the dominant electrical source, with wind turbines used as appropriate.

Nonelectric sources of energy will also be utilized. The primary space heating and hot water system will most likely be a ground-source heat pump powered using electricity from wind, solar, and stored energy. A direct solar collector could be also be used for hot water and water preheating to reduce energy usage by the ground-source heat pump. A combined heat and power generation system (biomass or hydrogen) could be used to further reduce heating and hot water load, as well as to reduce energy-storage requirements. Initially, fuel for the biomass plant will be wood and organic waste.

Wind and solar are presently the least expensive renewable energy sources per kilowatt-hour. We propose an installation of two to four turbines, moderately sized to reduce visual impact. Wind generators typically operate at a capacity factor of 20 percent to 40 percent, depending on local wind conditions, so a well-sited 1,000 kW wind setup would generate an average of about 300 kW (sometimes more, sometimes less) or 7,200 kWh/day.

Solar photovoltaics are now the least expensive new electricity source at many locations. Photovoltaic costs have been declining at about 3.2 percent per year since 1998, and are likely to continue this trend.[27] Solar generation typically operates at an annual capacity factor of 15 to 30 percent in the continental USA. A 1,000 kW solar array[28] might generate on average about 5,000 kWh/day. Thus 1,000 kW wind and 1,500 kW solar would generate the proposed annual average goal of about 15,000 kWh/day.

---

27   Galen Barbose, Naïm Darghouth, and Ryan Wiser, "Tracking the Sun III: The Installed Cost of Photovoltaics in the U.S. from 1998–2009," Lawrence Berkeley National Laboratory, December 2010, accessed May 21, 2012, http://eetd.lbl.gov/ea/emp/reports/lbnl-4121e.pdf.
28   For example, 1,000 kW-AC; 1,500 kW-DC; south-facing array, mostly unshaded, in the Northeast United States.

Solar's potential is gigantic: according to the US National Renewable Energy Laboratory,

> In the United States, cities and residences cover about 140 million acres of land. We could supply every kilowatt-hour of our nation's current electricity requirements simply by applying PV to 7 percent of this area—on roofs, on parking lots, along highway walls, on the sides of buildings, and in other dual-use scenarios.[29]

Shy rooftops nationwide are pleading to be covered by solar panels!

The sun will not only generate a chunk of our electricity, but it will also help heat our showers: a midsized active solar collector installation could be used to fully heat water in the summer and preheat it in the winter. A collector area of about 300 m^2 would be appropriate for Greenway. This technology is extremely mature: in 2010, solar hot water collectors worldwide were capable of producing 185 *million* kilowatts of thermal power.[30]

Biomass is sustainable, almost net zero as regards global-warming potential, and available locally. Small-scale burning of biomass, as in a woodstove, raises substantial concerns about emissions and byproducts (ash, smoke, and the like). However, if biomass is used to manufacture relatively clean-burning fuels at mid or large scale, substantial reductions in emissions and waste can be achieved. Sulfur dioxide emissions per unit energy for a high-temperature biomass gasification installation can be less than a tenth those of a home-based

---

29 "PV FAQs," National Renewable Energy Laboratory (US), February 2004, accessed May 21, 2012, http://www.nrel.gov/docs/fy04osti/35097.pdf.
30 "Renewables 2011, Global Status Report," Renewable Energy Policy Network for the 21st Century, August 2011, accessed May 21, 2012, http://www.ren21.net/Portals/97/documents/GSR/REN21_GSR2011.pdf.

woodstove or furnace; particulates, only 3 percent as much; carbon dioxide, one-sixth.[31]

A significant advantage of a biomass energy system is that it can run at any time—it is not reliant at any moment on whether the wind is blowing or sun shining—and can therefore be used to reduce reliance on energy storage. Biogas can produce electricity, heat, or both: in Europe, where hundreds of biogas systems are installed annually, it was already producing 4.9 percent of all electricity by 2019.[32]

By adjusting the mix of solar and wind, as well as possibly including hydro, geothermal, biomass, solar thermal, and other site-dependent alternative energy generation technologies, a totally green energy generation system can be designed for any geographic location in the United States.

Not Invited to the Party

Some alternative energy sources are in early stages of development or very site specific. Wave, tidal, and ocean-thermal power are relatively early in development, can be expensive, and require locating on coastal property. True geothermal heat would require favorable geology and deep drilling, neither of them likely options for Greenway. Nuclear fusion, satellite-based microwave power, and other speculative alternatives must find fulfillment elsewhere, as well.

Such omissions are OK because Greenway does not aim to do everything, to be everything: it aims to implement a richly diverse mixture of practical technologies that work for a particular real-world site. Since all actual sites have con-

---

31   Barbara Klingler, "Environmental Aspects of Biogas Technology," German Biogas Association, 1999, accessed May 21, 2012, http://homepage2. nifty.com/biogas/cnt/refdoc/whrefdoc/d7env.pdf.

32   Marc Schaller, "Biogas Electricity Production Hits 17 272 GWh a Year in Europe." Engineer Live, September 2008, accessed May 21, 2012, http:// www.engineerlive.com/Energy-Solutions/Waste-to-Energy/Biogas_electricity_production_hits_17_272GWh_a_year_in_Europe_/20788/.

straints, Greenway, by working within site-specific and scale-specific limitations, will actually be producing knowledge of universal value.

Conclusion

In this chapter, we have estimated the energy requirements of Greenway College and laid out the elements of a system that will meet the college's needs while also serving its teaching mission. Solar and wind electricity, ground-source heat pump, solar hot water—all these are in the picture, framed by the smart, efficient *use* of energy that defines Greenway's demand profile.

The technologies we favor are not only intrinsically sustainable but eminently practical: solar photovoltaics, wind, and solar hot water are well established, low maintenance, and reliable. Both solar and wind have recently become cost competitive with new fossil-fuel. These technologies produce intermittent output, but by combining these sources with other technologies, including energy storage—as reviewed in the next chapter—we can build a highly reliable, zero-sacrifice, sustainably powered stand-alone facility. At Greenway, we will engineer a beautiful energy machine that works 24–7–365.

Storing Energy

There must surely come a time when heat and power will be stored in unlimited quantities in every community, all gathered by natural forces. . . . When we learn how to store electricity, we will cease being apes ourselves; until then we are tailless orangutans. You see, we should utilize natural forces and thus get all of our power.

—Thomas A. Edison, 1910

Cellular phones are great for businesspeople on the go in the city, traveling salesmen, people like that, but coverage will never be cost-effective in rural areas and they'll never be able to compete against the huge infrastructure for landline phones. Do you know how much money has been put into running phone lines to every house in the United States? Sure, cellular phones will get smaller, better, and cheaper, but they'll never compete directly against landlines. Maybe in a few hundred years . . .

—Telecom engineer, to a
colleague in the early 1990s

Introduction

Energy is only useful if we can access it when, where, and how we want it. Chemical energy stored in combustible fuels like gasoline can't directly do many of the jobs that electrical energy can—you can't easily run a cell phone with a lump of coal. Electrical energy, on the other hand, is of the highest quality and can be harnessed to do almost any job, but it is expensive and bulky to bottle and carry around. We exclusively use large wires to bring electrical power to our buildings and homes; our technologies to carry around electrical energy with us and to move it into the future, to *when* and exactly *where* we want it, are more diverse—with huge amounts of energy stored for later use by pumping water uphill and tiny amounts of energy made portable in hearing aid batteries. Batteries and other methods of storing and carrying around energy tend to be more expensive than direct usage, but the need for portability in space and time often outweighs its costs: for instance, billions of batteries are sold each year in the US alone.

Combining zero-emissions sustainable energy generation (chapter 2) with technologies to bottle up useful energy and move it around space and time allows us to provide for all the energy uses of our society in a totally green, zero-sacrifice manner. The technologies that store some of the energy generated at Greenway, thus enabling the college to meet all its stationary and mobile energy needs—even when the wind falters or the sun goes behind a cloud—are the subject of this chapter.

Terms of Engagement

We utilize electrical, mechanical, thermal, and other forms of energy to run everything from hearing aids to delivery trucks to skyscrapers. No single technology can meet all these needs, so over the last century engineers have developed many sources of useful energy. From the point of view of portability, these energy sources can be divided into *stationary* and *mobile* types. Coal, nuclear, and hydroelectric power plants are stationary (except for a few antique steam trains and the occasional nuclear submarine). Some sources of energy can wear either hat: chemical fuels like diesel, methane, and hydrogen can be (and are) used to power both stationary and mobile machines. Finally, most chemical batteries, such as car batteries, flashlight batteries, or the rechargeable batteries found in personal electronics, are used in mobile devices. Some recent installations of chemical batteries paired with wind and solar chemical batteries—as big as trailer homes and many times heavier—are stationary.

Batteries and fluid fuels are the dominant mobile energy sources. To clearly understand their roles, we must bring in a second distinction—*primary versus secondary* energy sources. Petroleum, uranium, sunlight, and rivers running downhill are primary sources because they provide us with useful energy that we would not have otherwise. On the other hand, a charged NiCad battery, a wound-up spring, or a cylinder of compressed air must get its energy from somewhere else: it is secondary. Primary sources charge secondary sources. Some fluid fuels, like biodiesel, natural gas, or kerosene, are primary sources, while others, like hydrogen derived from water, are secondary, holding a "charge" of energy originally harvested somewhere else.

Mobile versus stationary is a way of thinking about storing and using energy, while *primary versus secondary* is a way of thinking about where energy ultimately comes from. A third consideration is energy *density*. Each pound or gallon of any mobile energy source, whether primary or secondary, must deliver sufficient energy (and deliver it quickly enough) to justify the hassle and cost of hauling it around. A car battery the size of a house would be useless no matter how cheap it was. Stationary energy sources, on the other hand, can often get away with low energy density: they may collect energy from diffuse primary sources, like the wind, and store it in comparatively bulky forms of storage, like elevated artificial lakes.

In general, secondary costs more than primary and mobile costs more than stationary. Electrical energy from a watch battery—which is mobile, secondary, and miniaturized to boot—costs about one hundred thousand times as much as energy from a wall socket.[1] Yet there is still a market for watch batteries because their mobility and tiny size justify their high cost per unit of energy. Moral: when shopping for an energy supply for a specific application, raw cost per kilowatt-hour is seldom the whole story.

Liquid fossil fuels, for example, are cheap, stable, high-density primary energy sources—in some ways, a great buy. That's why we buy them. Yet they are also polluting to extract and burn, geopolitically invidious, and finite in supply. Someday, perhaps soon, crude oil production will fall behind swelling global demand and stay there. Synthetic fluids refined from tar sands and coal may patch the supply-demand gap for a while, but their extraction destroys or disturbs vast landscapes and their lifecycle greenhouse emissions are even greater than those of coal and oil. Wouldn't it be nice if dense, mobile energy could be derived from inexhaustible sources like sun, wind, and sewage?

Fortunately, they can. Electric cars can be charged by windmills and solar cells, and biodiesel or ethanol can be manufactured from renewable feedstocks. Hydrogen can be produced by electrolysis, the electrical splitting of H_2O into H_2 and O_2, while methane (a.k.a. natural gas) can be made in biodigesters from plant or animal waste. There are other options, too. The bottom line is that fluid fuels and batteries can provide mobile, high-density energy from clean, affordable primary energy sources.

Which means that by judiciously using batteries and manufactured fuels, Greenway College can meet all its mobile-power needs without fossil fuels. And by storing large

---

1   The factor of 100,000 is based on $0.14/kWh for grid power versus $14,000/kWh for a watch battery. The former figure is from http://www.eia.doe.gov/electricity/epm/table5_6_b.html, the latter from A. B. Lovins et al., *Small Is Profitable: The Hidden Economic Benefits of Making Electrical Resources the Right Size*, Rocky Mountain Institute, Snowmass CO, 2002 (p. 96).

amounts of energy in stationary devices, it can assure itself an uninterruptible, 24-7 supply of high-quality electrical power even while getting its energy mostly from intermittent sources like sun and wind.

Let's look more closely at how this can be accomplished.

Storage Basics

Rule Number One of energy storage is that you never get as much energy back out as you put in. Storage therefore costs you twice, once for the hardware and once for all the energy that is lost in the storage-retrieval cycle. On these terms, storage sounds crazy—an elaborate scheme for wasting energy—yet it is in fact often economic or even essential. This is obvious for mobile devices like laptops and cars, but the virtues of large-scale, stationary energy storage are also being increasingly sung by engineers the world round. Storage devices big enough to power a college campus can justify their cost in money and energy.

There are several methods for storing energy in bulk: the most common is to pump a few billion gallons of water uphill and stash them behind a dam. Electricity drives the pumps that raise the water and is recovered later when the water runs downhill through turbines. From the grid's point of view, a pumped-storage reservoir is just a huge rechargeable battery. High-pressure air can be saved in caverns or tanks. Large conventional chemical batteries can be charged and discharged. Methane generated by bacteria from sewage, cornstalks, or other organic junk can be stored in tanks, or hydrogen produced by cracking H_2O can be saved as a super-cold liquid or gas under high pressure or in metallic sponges.

The most commonly mentioned role for stationary storage these days is supporting renewable energy. Solar cells and wind turbines are famous for generating power intermittently; that is, only when the sun shines or the wind blows. This intermittency is important—but not as important, on the national scale, as is often claimed.[2] Solar output matches pretty well the daily peak demand for power, which occurs around midday and a bit later. Wind generation is intermittent at different times in different parts of the country—the wind is always blowing somewhere—so power transmission from one area to another balances some of the effects of intermittency.

Demand (GW)

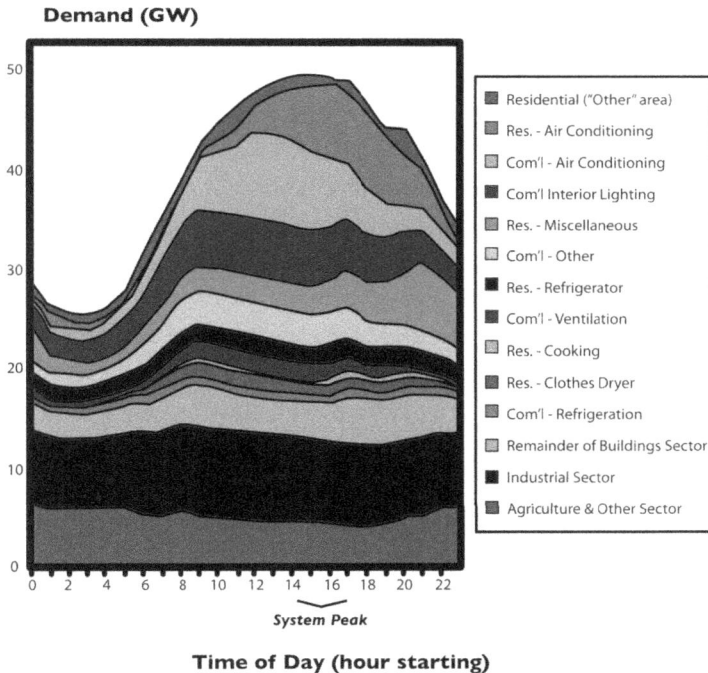

Time of Day (hour starting)

Fig. 3.1. California electric demand versus time of day for a hot summer day. Demand approximately doubled from its lowest point at 3 AM to its peak from 2 PM to 4 PM.[3]

2   Yih-huei Wan and Brian K. Parsons, "Factors Relevant to Utility Integration of Intermittent Renewable Technologies," NREL/TP-463-4953, August 1993, accessed July 23, 2012, http://www.nrel.gov/docs/legosti/old/4953.pdf.
3   Richard E. Brown and Jonathan G. Koomey, "Electricity Use in California: Past Trends and Present Usage Patterns," Ernest Orlando Lawrence Berkeley National Laboratory, May 2002, accessed June 6, 2012, http://en-

Another motive for purchasing storage is the fact that electrical *usage* is also not constant like wind or solar *supply*. As shown in figure 3.1, electrical demand is much higher during the daytime from about 8 AM to 9 PM. The top-of-the-curve "system peak" is highest during heat waves, when many air conditioners get turned on at once—or, in the South, during cold snaps, when many electric heaters get turned on. Peak demand can be met by turning on generators that burn fuels like natural gas, or by tapping other stored energy, or by some combination of the two.

Greenway Needs Storage

As explained in the previous chapter, Greenway won't need the grid and will be capable of operating independently from it. If grid connection is the usual method of getting reliable electrical power, how will Greenway ride out the dips and peaks of its solar and wind systems without it? The answer, of course, is storage. When the sun isn't shining or the wind dies down, Greenway's energy system will draw as much energy as it needs out of storage. This will happen automatically, without a hiccup: users will never notice the difference. Indeed, Greenway's stand-alone energy system will likely be *more* reliable than the grid, which fails often enough in the US to cause $57 billion of economic losses per year.[4] Also, Greenway must pack some of its renewable energy into secondary, mobile forms, namely batteries and fluid fuels, to run its portable equipment and vehicles.

The inherent trade-off of all energy storage—hardware costs and energy losses versus increased reliability and mobility—must be faced realistically by anyone seeking

duse.lbl.gov/info/LBNL-47992.pdf.
4   K. H. LaCommare and J. H. Eto, "Understanding the Cost of Power Interruptions to U.S. Electricity Consumers," Ernest Orlando Lawrence Berkeley National Laboratory, September 2004, accessed May 21, 2012, http://certs.lbl.gov/pdf/55718.pdf.

renewables-based energy independence. (Some alternative energy generation systems, such as geothermal or biomass, are not intermittent, but these are unlikely to supply as much electricity to Greenway as inherently intermittent solar and wind.) This investment of dollars and additional energy for the functionality gained by energy storage is not dissimilar to the investment in dollars and energy for petroleum extraction, processing to gasoline, shipping, and distribution. Factors in evaluating the practicality and affordability of any system for storing energy include storage efficiency and energy losses, expected lifespan, start-up costs, and maintenance costs. Others are the speed at which the energy can be recovered upon demand, the duration of time that the energy can be efficiently stored, and, especially for mobile applications, weight and volume considerations. Against all these costs stand the valuable services that storage can provide.

Several effective energy-storage technologies are considered mature, while other technologies continue to be improved. The latter include ultracapacitors, superconductors, compressed-air systems, and hydrogen fuel cells. All have start-up costs and are ultimately energy consumers but can help transform intermittent renewable power into reliable power. Below, we take a closer look at some technologies that might play a role at Greenway College, both for stationary and mobile energy supply.

Meet the Candidates

Pumped hydro storage presently accounts for 99 percent of existing grid-scale storage worldwide.[5] But the times they are a-changing. Storage reservoirs can be built only in certain places, are economic only if gigantic, take minutes (not seconds) to respond to demand, and have other drawbacks.

5   "Energy Storage: Packing Some Power," *The Economist*, May 3, 2012, accessed May 22, 2012, http://www.economist.com/node/21548495.

Table 3.1
Levelized cost of energy, lifecycle duration, and round-trip efficiency for a range of storage technologies

Storage Technology	Levelized Cost of Energy ($/kWh)	Lifecycle	Round-Trip Efficiency
Hydro storage	$$	75 years	70%
Classical compressed air (combusted with natural gas)	$$	40 years	35–65%*
Advanced compressed air (advanced adiabatic or isothermal)	$$	40 years	50–70%[†]
Hydrogen (electrolysis & combustion engine)	$$$	10 years	25%
Hydrogen (electrolysis & fuel cell)	$$$$	5 years	35%
Sodium sulfur batteries	$$$	10 years	60–70%
Flow batteries (e.g. vanadium redox, zinc bromide)	$$$$	8 years	60–70%
Lead acid batteries	$$$	3 years	75%
Li-ion batteries	$$$$	5 years	95%
Flywheel	$$$$$	25 years	95% (1% parasitic)
Ultracapacitor	$$$$$	25 years	97%

Sources for data include author estimates, the book EPRI-DOE Handbook of Energy Storage for Transmission and Distribution Applications, and World Energy Council Report, "Energy Storage Monitor: Latest Trends in energy storage | 2019," https://www.worldenergy.org/assets/downloads/ESM_Final_Report_05-Nov-2019.pdf.

Technologies are ordered from multiple-hour to short-term storage. Levelized cost of energy (LCOE) is the total lifetime cost of a storage system divided by the total amount of energy delivered by the system. Values are approximate, case dependent, and may change with technology advancement. No solid independent study of LCOE for storage is yet available, and thus relative values have been used where $$ implies < $0.20/kWh, $$$ <$0.50/kWh, $$$$ < $1.00/kWh, and $$$$$ > $1.00/kWh, with cost of fuel based on average nighttime U.S. electricity costs.

Notes: *There are various ways to calculate efficiency for compressed-air energy storage hybridized with natural gas combustion. (Samir Succar and R. H. Williams, "Compressed Air Energy Storage: Theory, Resources, and Applications for Wind Power," Energy System Analysis Group, Princeton University, April 8, 2008.) † No commercial systems exist to date for this technology.

Engineers have therefore struggled to develop more nimble alternatives, some of which are listed in the first column of table 3.1. The number of times a storage system can be charged and discharged before wearing out ("Lifecycle") and the fraction of energy stored that can be retrieved ("Round-Trip Efficiency") are also listed. (Round-trip efficiency can be interpreted as follows: If it takes 3 watt-hours of energy to charge a rechargeable AA battery that can supply only 2.5 watt-hours once charged, then the battery's round-trip efficiency is 2.5 divided by 3, namely 83 percent.)

The array of choices in table 3.1 may be at first sight bewildering. Should Greenway College buy pumped hydro storage or Li-ion batteries? Or advanced compressed air? Or maybe some kind of combo platter? Juggling all the factors and technological trade-offs to design Greenway's energy system is indeed a nontrivial job. However, if building a sustainable energy economy were as simple as plugging in a coffeemaker, we wouldn't need a Greenway College at all.

Below, we discuss and compare a few of the leading contenders for Greenway College's stationary (non-vehicular) storage needs, beginning with the most traditional: pumped hydro.

Pumped Hydro

As already noted, pumped hydro is presently the global champ among stationary energy-storage options. It stands out for its long lifetime (about seventy-five years), good round-trip efficiency (approximately 70 percent), and low cost per unit of energy stored and retrieved. That is why pumped hydro accounts for over 97 percent of grid energy storage in the US, with 23 GW.[6] The idea is simple: Electrical

---

6   National HydroPower Assocation. *2018 Pumped Storage Report*. Washington, DC. https://www.hydro.org/wp-content/uploads/2018/04/2018-NHA-Pumped-Storage-Report.pdf.

energy is used to pump water to an elevated reservoir (figure 3.2). The energy can be retrieved (mostly) when the water is run downhill through turbines. Pumped-hydro storage uses proven machine designs that have high efficiency, long lifetime, and low maintenance, but its high- and low-level reservoirs can require radical modification of the landscape.

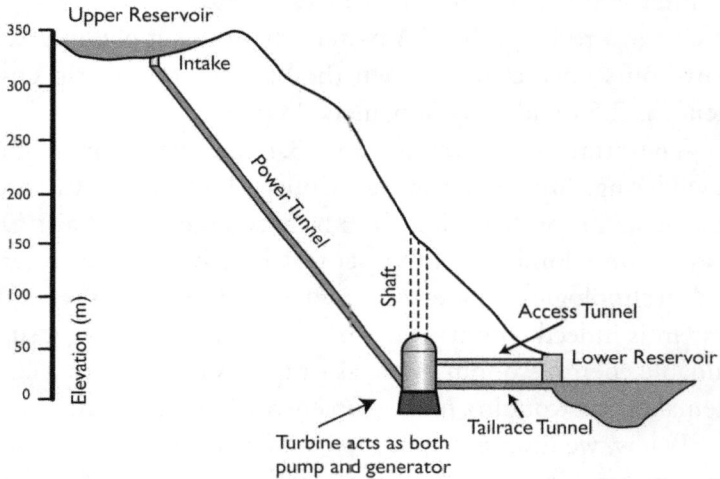

Fig. 3.2. Schematic of pumped-hydro energy storage. Electric power is used to pump water to the upper reservoir, usually at night, and is then generated by running water downhill again, usually during the day.

A key factor in the effectiveness of any pumped-hydro storage installation is its "head," the elevation difference between the upper and lower reservoirs. The higher the altitude to which a gallon of water is pumped, the more energy it can deliver when run downhill: doubling the head doubles the amount of energy stored by each gallon of water in the upper reservoir. Therefore, most pumped-hydro reservoirs are built on land at least a hundred meters above the lower reservoir.

Suitable sites for constructing large, high-head storage reservoirs are not to be found just anywhere. Fortunately, Greenway would require a reservoir much smaller than those built to service the grid, and consequently would find it easier to site its system. A fairly small upper reservoir—about 8,000 square meters area (2 acres), average depth 10 meters—would provide ample storage for the college's needs. For a reservoir of this size, sites with suitable elevation are not hard to find in the Northeast United States: at a closed-down Vermont college that appeared on the market, suitable for a Greenway campus, an upper reservoir could have been constructed with an appropriate 150 meters of head.

In sum, if Greenway College is built on a rural site with appropriate topography, pumped storage will be a strong contender for its primary energy-storage technology.

Compressed Air

Compressed-air energy storage (CAES) stores energy using the springlike compressibility of air. First, electricity turns a compressor to force high-pressure air into a cylinder or cavern. To recover this energy, the stored air is used to drive a generator. In existing utility-scale CAES systems, the pressurized air does not push a turbine by itself but is used to boost a natural-gas turbine that combines gas with compressed air before burning it. By using low-cost, off-peak electrical energy to generate and store compressed air, the amount of natural gas consumed to generate a unit of electrical energy is reduced the next day.

It's a good trick, but Greenway College won't be building a compressed-air-assisted natural gas plant. Even if we were willing to burn fossil fuels, which we're not—or able to generate enough renewable methane to run such a turbine— a Greenway-sized gas turbine would not be efficient: below about 2,000 kW, cost doubles and efficiency drops by about a third.[7] Greenway College, even during peak demand, will

---

7   Meherwan P. Boyce, *Gas Turbine Engineering Handbook*, (Burlington, MA: Gulf Professional Publishing, 2006).

need less than 1,000 kW of power (which is about as much as it takes to run a Super Wal-Mart).[8]

However, other forms of CAES may work for Greenway. Technologies that use compressed air alone as an energy-storage medium, with no need for the natural gas, may be scaled to campus size at good efficiency.

The biggest barrier to making CAES efficient and affordable is the drastic temperature change that can occur when air is compressed and expanded. Whenever air (or any gas) is compressed, it tends to become hotter. If compressed rapidly to many times atmospheric pressure, it can become hotter than any household oven—700 ºC (1,300 ºF) for air compressed from atmospheric pressure to 70 atmospheres, more than hot enough to melt aluminum. In a CAES system, all this heat—which represents much of the energy invested in the compression process—must either be kept from leaking away while the air is stored, which entails an insulated, more costly storage system, or it must be allowed to leak away. But if the inconvenient heat is simply discarded, that is a loss of energy and thus of storage efficiency. On the other hand, if the compressed gas is stored at room temperature, a new difficulty appears during recovery of energy—*frigid* temperatures. Gas tends to cool when it expands, so rapid expansion from high pressure can produce super-cold air (e.g., negative 200 ºC (–330 ºF) for air expanded from 70 atmospheres to atmospheric pressure, almost cold enough to liquefy the expanded air). Such extreme cold would force the use of expensive, cold-resistant materials, cause frost and ice buildup, and so forth.

Several CAES methods have been proposed that will deal with this issue. All involve moving thermal energy around to

---

8   Matthew L. Wald, "A Wind Farm Would Link Northeastern Grids," *New York Times*, December 9, 2010, accessed May 24, 2012, http://green.blogs. nytimes.com/2010/12/09/wind-farm-would-link-northeastern-grids/.

avoid inconvenient extremes of hot and cold. One can, for example, remove heat from the air during compression, store it (in a tank of hot water or some other heat reservoir), and restore it to the air during expansion.

In any form, CAES is pumped hydro's closest competitor for convenience, safety, and cost. Although midscale, aboveground, air-only CAES technologies are not yet established in the market, they may be soon. They will offer several advantages: first, their air tanks and other components can be located most anywhere, while a pumped-hydro storage reservoir cannot. Second, CAES with aboveground air storage will not impact the local ecology and hydrology as much as reservoir construction does. Third, such technologies are likely to become commonplace in the grid in coming years, and it is precisely such technologies that Greenway College students and faculty must investigate, understand, challenge, and improve.

Thermal Energy Storage

Thermal energy is the random motion of atoms and molecules jiggling in place or flying about. Because thermal energy is disorganized, it is more difficult to harness in the form of useful work than is the highly ordered energy of a charged battery, an elevated body of water, or a spinning flywheel. Nevertheless, because thermal energy is easy to produce from sunlight or fire, we use thermal systems for most electrical energy generation (making steam in power plants) as well as for direct heating and cooling.[9]

Out-of-place, excess thermal energy is actually a burden rather than a resource. When fuel burns in the cylinder of a car engine, it is generating thermal energy, which causes the

---

9   Though in colloquial speech "heat" and "thermal energy" are used interchangeably, in physics "heat" strictly means energy transferred between one system and another by a thermal process.

gas trapped in the cylinder to expand, pushing on the piston and ultimately turning the wheels. Some of the thermal energy from combustion is thus converted to mechanical work, but most cannot be put to use and must be thrown away by the radiator. In stationary settings, we can be more clever and efficient: "combined heat and power" systems harness otherwise wasted heat from electric generation (in, say, a wood or gas-fired plant) to heat buildings, water, or the like. Another option is to *store* thermal energy for later use: heating, cooling, or energy generation.

Changing phase, for example from solid to liquid (as in freezing water), or liquid to gas, is also a form of thermal energy storage, because a phase change is accompanied by a change in the energy content of substance with little or no change in temperature. Water actually releases thermal energy as it freezes—0.1 kWh per kilogram—which is why orange growers spray water on their trees to save their fruit during a frost.[10] Some thermal-energy storage systems rely on a substance that undergoes a phase change.

Solar thermal-energy storage systems heat substances during the day, when solar energy is available, and then use the stored energy as needed, day or night. The simplest and most widespread form of thermal storage is residential hot water. Seasonal hot-water storage has also been implemented, as in the central hot-water heating system that has served several hundred apartments in Friedrichshafen, Germany, since 1996.[11]

At large scale and high temperatures—high enough to

---

10   Arlie Powell, "Methods of Freeze Protection for Fruit Crops," Alabama Cooperative Extension System, October 2000, accessed May 22, 2012, http://www.aces.edu/pubs/docs/A/ANR-1057-B/index2.tmpl.
11   Thomas Schmidt and Janet Nußbicker, "Monitoring Results from German Central Solar Heating Plants with Seasonal Storage," ISES 2005 Solar World Congress, Orlando, Florida, August 6–12, 2005, accessed May 22, 2012 At http://www.solites.de/download/literatur/05-03.pdf.

melt salt—thermal storage allows some solar electric-generating installations, such as the Gemasolar thermal-solar plant in Andalucía, Spain, to supply electricity long after the sun has set. The Gemasolar plant (operational since 2011) stores enough solar-heated molten salt to provide power during fifteen hours of darkness.[12]

At Greenway College, thermal energy storage will at first be used mainly to supply hot water—an important role, as hot water constitutes about 15 percent of all energy use in residential buildings.[13] Self-heating, self-cooling buildings, such as are described in the next chapter, may also use internal thermal storage to help keep themselves at a steady, comfortable temperature.

A thermal-solar electric generating station, with or without thermal storage, is currently not proposed for Greenway. Other storage technologies that will probably not be considered for reasons of scale and safety include large stationary batteries based on flowing chemicals ("flow batteries") or batteries based on high-temperature sodium and sulfur, although these may see increasing application in the grid.

Hydrogen

Hydrogen is a promising secondary energy carrier, especially for transportation and portable devices. It is one of the densest fuels in terms of energy stored per unit weight, and when it is combusted or reacted with oxygen the result is chiefly water. Although most hydrogen is manufactured by combining natural gas with steam, hydrogen (H_2) can also be

---

12   "Gemasolar Thermosolar Plant," National Renewable Energy Laboratory, 2011, accessed May 22, 2012, http://www.nrel.gov/csp/solarpaces/project_detail.cfm/projectID=40.

13   "High Efficiency Water Heaters," US Environmental Protection Agency, 2006, accessed July 24, 2012, http://www.energystar.gov/ia/new_homes/features/WaterHtrs_062906.pdf.

produced by cracking water molecules (H_2O). This cracking process consumes energy: when the hydrogen recombines with oxygen, the energy is released. Recombination of hydrogen with oxygen may occur with high efficiency in a fuel cell, a device that produces electricity directly from chemical reactions. Hydrogen can also be burned in internal combustion engines or turbines. Its advantages include its ability to double as a vehicular fuel or stationary storage medium; even cell phones and laptops can be powered by hydrogen reacted in miniature fuel cells.[14]

A clean, elegant cycle thus beckons: water plus electricity makes hydrogen, hydrogen plus oxygen makes water and electricity. One can imagine an all-electric campus or entire society where electricity from solar cells and windmills makes hydrogen from water, and the hydrogen runs vehicles or produces electricity in fuel cells.

There are obstacles, however. Although a kilogram of hydrogen holds a lot of energy, it also takes up a lot of space: at any given temperature and pressure it is the least dense of all materials. High pressures, low temperatures, or sophisticated metallic sponges are required to store large quantities of it in confined spaces. This raises costs, especially for vehicular uses. Also, round-trip efficiency for a water-hydrogen-water cycle is relatively low: presently less than 35 percent, versus about 70 percent for pumped hydro or chemical batteries or 50 to 60 percent for compressed air.

A study by the US National Renewable Energy Laboratory found that a fuel-cell system deriving its hydrogen from water could probably compete economically with batteries

---

14   Lisa Zyga, "Hydrogen-Powered Cell Phone Doubles Battery Lifetime," PhysOrg, January 14, 2008, accessed May 24, 2012, http://www.physorg.com/news119544735.html. Also, Carol Potera, "Beyond Batteries," *Environmental Health Perspectives* 115:1 (2007): A38–A41, accessed May 24, 2012, http://www.ncbi.nlm.nih.gov/pmc/articles/PMC1797861/pdf/ehp0115-a00038.pdf.

for energy storage in installations of middling size.[15] And despite a popular belief that hydrogen is uniquely dangerous, it is not: leaked hydrogen disperses rapidly, unlike gasoline, and burns with a pale flame that radiates little heat and so does not readily ignite nearby objects.[16] Safety studies have shown that hydrogen-powered vehicles can be at least as safe as conventionally fueled vehicles.[17] Hydrogen-powered cars exist today only in very limited market offerings, but modification of four-cycle internal combustion engines to run on hydrogen is straightforward (although in such engines, proper fuel-to-air ratio is key to minimizing production of nitrous oxide, a pollutant).

All things considered, it is highly likely that Greenway will use hydrogen for both long-term stationary energy storage and for some of its maintenance equipment and vehicles. Such technologies are already on the market: for example, Hydrogenics Corporation (www.hydrogenics.com) offers a complete lineup of components for hydrogen power backup, including electrolysis, compression, storage, and generation. For approximately $1 million, a system can be purchased that generates 4 kg/hour of hydrogen from water and stores it at 400 atmospheres, including a dispenser for mobile uses and a 125 kW stationary backup generator. This system can

15   Darlene M. Steward, "Analysis of Hydrogen and Competing Technologies for Utility-Scale Energy Storage," NREL/PR-560-47547, US National Renewable Energy Laboratory, February 11, 2010, accessed May 23, 2012, http://www.nrel.gov/docs/fy10osti/47547.pdf.

16   "Fuel Cells Technologies Program," US Department of Energy, February 2011, accessed May 22, 2012, http://www.hydrogen.energy.gov/pdfs/doe_h2_safety.pdf.

17   "Analysis of Published Hydrogen Vehicle Safety Research," DOT HS 811 267, US Department of Transportation, National Highway Traffic Safety Administration, February 2010, accessed May 22, 2012, http://www.nhtsa.gov/DOT/NHTSA/NVS/Crashworthiness/Alternative%20Energy%20Vehicle%20Systems%20Safety%20Research/811267.pdf.

generate power from its stored hydrogen for approximately twenty-four hours.

Fast Responders

So far we have been speaking of bulk energy storage—devices that can meet the energy needs of a whole campus for many hours on end, if need be. The technologies that supply such storage tend to be slow to respond, however, compared to the speed at which a computer screen goes blank when the power flickers. To produce truly uninterruptible power, to fully smooth the continually twitching mismatch between supply and demand, faster-acting forms of storage are needed: batteries, flywheels, ultracapacitors, and superconductors. These tend to be expensive for storing bulk energy, but such technologies have fast response times and good power density.

Batteries. Electrochemical batteries, which include such well-known products as alkaline AA batteries and lead-acid car batteries, are the most mature technology for meeting short-term energy demand. They have disadvantages, however. Lead-acid batteries offer low cost, but moderate energy density and reduced lifetimes. Lithium-based technologies offer higher energy density and longer lifetime, but traditionally at high cost. Lithium-based batteries have been coming down rapidly in price with widespread usage in electric vehicles and stationary storage. A host of other battery chemistries have been developed, each filling some market niche and having its own advantages and disadvantages. Greenway College will study all these technologies and work with many of them, especially for transportation and short-term storage, but will also rely on some of the long lifetime alternatives described above for bulk storage.

Flywheels. Flywheels can be used to store electrical energy as kinetic energy—the energy of moving matter—in a heavy, rotating disk (a "flywheel.") With very low-friction bearings, such as floating the disk in a vacuum with magnets, energy can be stored with little loss for long periods of time.

The amount of energy stored is determined by a flywheel's mass, shape, and speed. To store more energy in a flywheel, one can either build a bigger flywheel or a denser flywheel, or spin the flywheel at higher speeds.

Flywheel storage is commercially available today for uninterruptable power supplies and, incipiently, for maintaining the quality of power supplied by the grid. For example, Active Power of Austin, Texas, offers an energy-storage system using a notched steel flywheel that is capable of storing 1.8 kWh of energy and delivering it at rates of up to 500 kW. That is enough energy for a cluster of computers to ride out a brief interruption in grid power. In 2011, an installation of two hundred much bigger flywheels built by Beacon Power began operation in Stephentown, New York. The flywheels are charged (spun up) during off-peak hours, then tapped for power whenever spiking loads threaten to drag down the frequency of the AC power on the grid. Each Beacon unit is about 10 feet tall, spins a magnetically levitated flywheel at 16,000 revolutions per minute in vacuum, and stores up to 25 kWh.

For Greenway College, units like the Active Power system might provide a long-lifetime, low-impact system for rapid response.

Ultracapacitors. Electric charges can be held apart from each other in devices called capacitors, storing energy. Tiny capacitors are found in most all electronic devices, and large ones are contemplated for fast-response energy storage.

Since capacitors store their energy directly as separated electrons, they can start delivering that energy in thousandths or even millionths of a second. They can be reused thousands or millions of times, while electrochemical batteries can be charged and discharged a few thousand times at most. They can also be charged more quickly than any other storage device and have no moving parts—always a plus.

Capacitors' biggest drawback is that they do not store much energy for their size (low energy density). Efforts to fix this problem have led to the development of "ultracapacitors" or "supercapacitors," which can store hundreds of times more electrical energy than similarly sized traditional capacitors,[18] which is a third to a half as much as a similarly sized lead-acid batteries. Ultracapacitors cannot deliver their energy quite as quickly as traditional capacitors, but are still about ten times faster than a chemical battery with similar capacity. Plenty fast.

Cost remains high, about ten to one hundred times higher than for an equivalent amount of lead-acid battery storage. Nevertheless, because an ultracapacitor lasts so long, the *lifetime* cost of storing a unit of energy can be as much as ten times less for an ultracapacitor than for a chemical battery.

Large ultracapacitors have been used in some buses, trains, and submarines, but are not yet at widespread work in the grid due to their high up-front cost. Greenway will keep a close eye on this technology as it evolves.

Superconducting Magnetic Energy Storage. Energy can be stored in a superconductor, which is any substance that presents zero resistance to the flow of electric current. Like capacitors, superconducting storage devices can yield their energy at electronic speeds. However, they currently only work if very cold: for superconductors, liquid nitrogen at about 200 °C below zero (–330 °F) is considered shockingly warm. Superconductors with their zero resistive losses are attractive to the zero-waste ethos of Greenway College, but

---

18   Jason Lee, "Ultracapacitor Applications for Uninterruptible Power Supplies (UPS)." White paper, Maxwell Technologies, undated, accessed May 23, 2012, http://www.maxwell.com/products/ultracapacitors/docs/201202_whitepaper_application_for_ups.pdf.

due to cost and availability, superconductor energy storage will likely be limited to laboratory study at the outset.

Taking Storage on the Road

Greenway will maintain a fleet of vehicles—its own vehicles, that is, not the hodgepodge of student-owned cars generally found in any college's most remote parking lots. (Perhaps a campus-run car-sharing system modeled on Zipcar can induce green-minded students to leave their personal automobiles, if any, at home.) Fueling vehicles can seriously dent a college budget: in 2010, Ohio State burned over two million liters (over half a million gallons) of vehicular fuel at a cost of almost $800,000.[19]

As mentioned earlier, energy density is crucial for mobile energy sources. Figure 3.3 shows that the energy density of all fluid and gaseous fuels is considerably higher than current battery technology. Yet batteries can be charged directly from solar panels and wind turbines, without the bother and inefficiency of manufacturing fluids. Should Greenway use fluid fuels, batteries, or some blend?

Of course, the status quo method of running our plows, staff cars, pickup trucks, lawn mowers, and other vehicles would be to use old-fashioned gasoline and diesel, like any normal college. Instead, we strive to make that status quo seem as strange and wasteful as wood-fired locomotives and gas lighting. There are clean and efficient ways to run vehicles, maintenance equipment, and portable electronics—ways that involve no sacrifice in performance, enhance energy independence, and are clean and affordable.

---

19   Gordon Gantt, "Ohio State Must Fuel Fleet at All Costs," *The Lantern*, May 18, 2011, accessed May 23, 2012, http://www.thelantern.com/campus/ohio-state-must-fuel-fleet-at-all-costs-1.2229787.

Energy per volume for mobile storage technologies

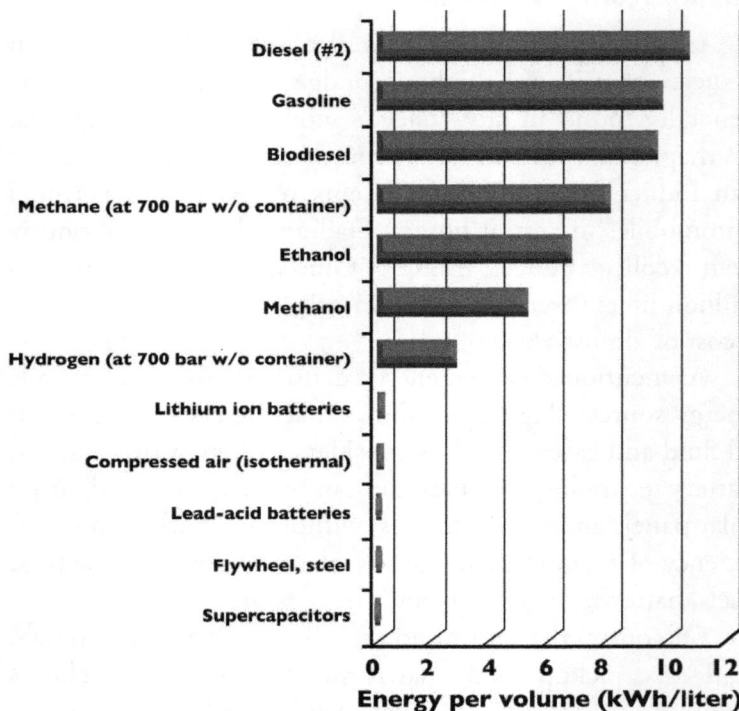

Fig. 3.3. Approximate energy densities of various storage materials, defined here as the amount of energy stored in one liter (about a quart) of the given medium.[20]

---

20   The numbers in this table have been assembled from a variety of sources, including author estimates, the website of the Electricity Storage Association (http://energystorage.org/tech/technologies_comparisons.htm), the US DOE's "Properties of Fuels Table" (http://www.afdc.energy.gov/afdc/pdfs/fueltable.pdf), and the following articles: A. Rufer, "Solutions for Storage of Electrical Energy, " EPFL, proceedings of ANAE in 2003, and Richard F. Post, T. Kenneth Fowler, and Stephen F. Post, "A High-Efficiency Electrome-chanical Battery," *Proceedings of the IEEE* 81:3 (1993): 462–474.

All-Electric Vehicles

Greenway will produce all its own electricity and charge electric vehicles primarily during times of excess generation. Electric vehicles have been intensively researched for decades. They are quiet, reliable, and as clean as the electric generators that charge them. The lithium ion batteries (and to a lesser extent nickel-metal hydride) used in today's electric and hybrid cars are a technologically mature form of energy storage that will certainly see use on Greenway campus.

A longstanding drawback of electric vehicles has been their limited range, which until recently was a fraction of gasoline-powered vehicles'. For all but the longest distance driving, electric vehicles and equipment now perform superior to internal combustion engine vehicles and equipment. Electric vehicles and equipment have truly reached the age of zero sacrifice.

A remaining drawback of electric vehicles is recharge time: a conventional car can gas up in a minute or two, but an electric vehicle can take hours to recharge. This is not much of a disadvantage for Greenway, however. Because the college will own enough vehicles and equipment, charging can happen when convenient and balance electricity usage.

Ethanol, Methanol, and Biodiesel

Other options for mobile energy storage include carbon-based secondary (synthetic) fuels such as methanol, ethanol, biodiesel, and methane, all of which have decent energy density. They can be made from biomass feedstocks such as wood, corn, soy, and organic waste and burned in conventional engines. In some cases they can be reacted in fuel cells.

The environmental impacts of these fuels vary widely, depending on how they are made: corn-based ethanol, found in gasoline across the US, has been said by some analysts to

consume nearly as much (or even more) energy in its manufacture than it yields when burned.[21] Even optimistic analyses
indicate that for each unit of energy released by combust-
ing corn ethanol, three-quarters of a unit of energy must be
invested to grow and process the corn. Nor does an energy-in,
energy-out analysis take into account soil loss from conventional corn growing, nutrient pollution from fertilizer, and
higher food prices worldwide due to the diversion of arable
land to ethanol production.[22] The "greenness" of corn ethanol
is therefore questionable.

At the other extreme, methane gas is spontaneously
emitted by bacteria digesting sewage or landfill waste. If this
methane is not collected or burned, but dispersed directly to
the atmosphere, it causes about twenty times as much greenhouse warming, molecule for molecule, as carbon dioxide.
For this reason, landfill methane is sometimes flamed off (in
effect turning it into carbon dioxide) as a greenhouse mitigation measure. Using it as a fuel is therefore a double win:
useful energy plus greenhouse mitigation. New York City
already captures half the methane emitted from its sewage
plants and burns it to meet 20 percent of the energy needs of
its wastewater treatment system.[23]

Greenway may use biomethane from digesters on a small
scale and will undertake research projects on biofuels.

---

21   D. Pimentel and T. Patzek, "Ethanol Production Using Corn, Switch-
grass, and Wood; Biodiesel Production Using Soybean and Sunflower,"
Natural Resources Research 14:1 (2005): 65–76. But see also http://www.
transportation.anl.gov/pdfs/AF/265.pdf (and criticisms of the latter at http://
www.consumerenergyreport.com/2006/03/30/how-reliable-are-those-usda-
ethanol-studies/).

22   Elizabeth Rosenthal, "U.N. Says Biofuel Subsidies Raise Food Bill and
Hunger," *New York Times*, October 7, 2008, accessed May 22, 2012, http://
www.nytimes.com/2008/10/08/world/europe/08italy.html.

23   Mireya Navarro, "City Is Looking at Sewage Treatment As a Source of
Energy," *New York Times*, February 8, 2011, accessed May 23, 2012, http://
www.nytimes.com/2011/02/09/science/09sewage.html?_r=1.

Energizing Greenway

Greenway will provide amenities equal to or better than those of a conventionally powered campus, all while having neutral or positive impact on the environment and using no fossil fuels. Thus, in our first implementations of energy generation and energy storage, we will rely on tried and true technologies that entail no sacrifice of reliability and thus of comfort. More experimental systems will not be implemented until after the successful establishment of the initial college buildings and their supporting technology.

Depending on the pace of technology development, this may make small-scale pumped-hydro storage our on-campus bulk storage technology of choice. Small-scale pumped hydro would not involve jumping any large technological hurdles, since its piece parts—pipes, turbines, pumps, and generators—are highly reliable. However, a campus having fairly high relief, with sites suitable for the construction of upper and lower reservoirs, would be needed.

Relatively well-established, long-term storage options that do *not* depend on site topography include chemical batteries, compressed air, and hydrogen. We will create hydrogen as fuel for vehicles and also perhaps for stationary backup power. The low round-trip efficiency of hydrogen makes it less attractive as our sole bulk-storage solution. Biofuels may supplement our stored energy supply at reasonable cost: for instance, methane from biodigestion is a portable, storable source of primary energy. Finally, high-speed, low-capacity energy storage will also probably be installed, based on flywheels, ultracapacitors, or batteries, to render small glitches in the on-campus power supply invisible to users.

There is, if anything, an embarrassing wealth of options. Technologies are developing fast, and all options must be weighed at the time of initial campus design. The best buys must be installed in such a way as to not foreclose future evolution of Greenway's energy system. The goal is not to supply Greenway with a perfect, finished energy system on day one, even if that were possible, but to supply it with an *evolvable* energy system that serves the campus's needs with high reliability while allowing students and staff to tinker, test, research, improve, reform. But the time to act is now.

Zero Waste

In our workshops we pride ourselves on discovering a use for what had been previously regarded as waste, but how partial and accidental our economy compared with nature's. In nature nothing is wasted. Every decayed leaf and twig and fiber is only the better fitted to serve in some other department, and all at last are gathered in her compost heap.

—Henry Thoreau, *Journal*, January 13, 1856

You [shouldn't just] filter smokestacks or water. Instead, you [should] put the filter in your head and design the problem out of existence.

—William McDonough, *Cradle to Cradle*

Zero waste is a goal now being seriously discussed by companies, universities, governments, and other institutions around the world. On the surface, it is simple: emulate nature by either returning all used materials to nature *or* repurposing them without causing net harm to people or the environment. This means no landfilling; no spewing smoke, toxins, or greenhouse gases; leaving water as clean as or cleaner than one found it; total recycling, reforming, or repurposing of all sewage, food leftovers, paper, metal, plastic, glass, and other materials; and high-efficiency products and processes of all kinds. Admittedly, zero waste is a visionary goal, yet a growing technical literature, backed by real-world experience, shows it to be practical.

At Greenway we will aim not at marginal, halfhearted improvements, but at the total elimination of waste. We will design our campus and activities to result in no harmful effluent, no solid waste, and no harmful emissions, while maximizing energy efficiency. In addition, we will seek to design, develop, test, and promote systems and initiatives to allow industries, homes, and other stakeholders to operate with zero waste at zero sacrifice. That last term is, as we have emphasized in previous chapters, crucial: any zero-waste program that asks people to live in sacrificial misery is bound to be a nonstarter. When we say zero waste, we emphatically mean *zero waste at zero sacrifice.*

Questions

What, exactly, do we mean by "waste" and "zero"?

Waste may broadly be defined as anything that is thrown away. Some advocates of zero waste advocate replacing the idea of "waste" with that of *residual product* or *potential resource.* These terms highlight the fact that in a perfectly rational industrial system, maximum value would be extracted from 100 percent of resource.

"Waste" is not only a lost opportunity, but is often positively harmful, in which case we call it "pollution." Indeed, since all resources require energy for their extraction and processing, and all energy production—even from renewable sources—entails some degree of pollution or other environmental harm, there is no such thing as truly harmless waste.

As for the "zero" part of "zero waste," it is not as simple as it sounds. Two basic senses or limiting cases of "zero" waste can be defined, which we might call *everywhere zero* and *net zero.* An *everywhere-zero* system produces no waste at any point. For example, it produces no combustion products anywhere—any of its exhausts could be routed safely into a child's bedroom. A *net-zero* system, on the other hand, may produce waste outputs at some points, but makes up for them at other points. If it burns fuel in one place, it creates fuel in another, perhaps by biodigesting food and sewage to yield methane, or by growing trees or other biofuel feedstocks, or by using wind or solar electricity to make hydrogen.

A net-zero system might be easier to build, but an everywhere-zero system might be easier to keep honest. Strict accounting is needed to assure that a system is truly net-zero, and apples-to-oranges comparisons may be unavoidable. Carbon offset or credit systems, which attempt to cancel the net emissions of far-flung enterprises, have been notoriously hard to verify.[1]

On the other hand, strictly net-zero systems can actually be built, while strictly everywhere-zero systems cannot. To take just one example, each of us exhales carbon dioxide, a greenhouse gas, and we prefer to not stop breathing. Our CO_2 output is trivial, but not *zero*.[2]

Greenway will aim at "zero," however defined, because all waste is an opportunity turned into a burden. Wasted material is discarded resource, and discarding resources is economically irrational; and on a finite Earth, zero waste is the only standard that we can uphold indefinitely. Mines will eventually play out, gas or oil wells will run dry, arable soil will wash to the sea, if we continue on our current one-way path: the timing is debatable but the outcome is inevitable.

And in the long run, why settle for zero? Why should not Greenway ultimately be a *negative*-waste system, producing more clean water, energy, and food than it takes in, with an

---

1   Mark Schapiro, "Conning the Climate: Inside the Carbon-Trading Shell Game," *Harper's*, February 2010, accessed May 23, 2012, http://citizensclimatelobby.org/files/Conning-the-Climate.pdf.
2   For a glimpse of just how technical a discussion of "zero" can get, see NREL on definitions of net zero energy use by buildings: P. Torcellini, et al., "Zero Energy Buildings: A Critical Look at the Definition," NREL/CP-550-39833, August 2006, accessed May 24, 2012, http://www.nrel.gov/docs/fy06osti/39833.pdf.

even greater margin of profit and pleasure? (To begin with, however, Greenway will stick to the goal of zero waste.)

Zero waste is both a vision and a collection of techniques. Below, we review the vision more closely, then touch upon some of the technologies that put it within reach.

The Vision

An ideal zero-waste industrial society would design and produce cars, houses, computers, and other products that optimize usage of resources during construction, life, and end of life. It would build things elegantly. It would "precycle" all its products, engineering them for easy breakdown when the time comes to reuse their ingredients. It would recycle all solid materials throughout the product lifecycle without degradation or "downcycling" to lower-quality products, reducing landfill and incinerator burdens to zero. It would produce zero pollution and thus suffer zero pollution-related disease. (Over 750 thousand people die every year, world-wide, from outdoor air pollution alone, and millions more from indoor air pollution.[3]) Its buildings would collectively produce their own heat, cooling, light, and power and would separate and treat their own graywater, sewage, and food left-overs. By using zero grid energy on-site, such buildings would incur zero emissions at distant power plants. Such a society would be *more* comfortable than ours, and its thrift would be technologically built-in and voluntary, not imposed on uncomfortable, unwilling citizens. And it could cost no more than, or less than, today's business-as-usual system.

Such a vision is so at odds with our current economy that it may seem a utopian pipe dream. But every single techni-

---

3   "Pollution: Costs of Inaction," *OECD Observer* 263, October 2007, accessed May 23, 2012, http://www.oecdobserver.org/news/fullstory.php/aid/2351/. Also, "Indoor Air Pollution and Health," World Health Organization, Fact Sheet No. 292, September 2011, accessed May 23, 2012, http://www.who.int/mediacentre/factsheets/fs292/en/.

cal building block of such a system has already been demonstrated. It is not technological inability that keeps us mired in our current puddle of cost, waste, discomfort, and pollution, but cultural inertia, irrational regulation, and simple lack of knowledge.

The economic and health costs of waste are heavy. Many of those costs have traditionally been externalized, that is, ignored by the waster because somebody else, downstream or downwind, pays the "cost" in discomfort or disease or rising sea levels or the like. Solid-waste disposal costs are not as easily externalized as those of other types of pollution, which is why cities, colleges, and other institutions must pay through the nose ($30 to $100 per ton) to have their solid waste hauled to landfills. Landfills occupy expensive real estate, can pollute groundwater, and are the USA's largest source of methane, a greenhouse gas twenty times more potent, molecule for molecule, than carbon dioxide. Incinerators are one alternative to landfilling; they can be integrated with recycling of noncombustibles such as metals and can generate heat and electricity. All existing incinerators pollute the air to some extent, however, especially since they have almost no control over the makeup of their incoming fuel (i.e., garbage). Today's trash-to-power facilities emit more CO_2 than a coal-fired power plant per unit of electricity produced.[4]

Perhaps a more explicit term than "zero waste," though clumsier, would be *zero pollution, zero hazardous waste, zero landfill*. In such a system, all emissions and unused hazardous and nonhazardous materials are captured during manufacturing or routine operations. These captured materials are then reused or processed and may be sold to other manufacturers (e.g., as in Rent-a-Solvent schemes). The final products them-

---

4   "Air Emissions," US Environmental Protection Agency, accessed March 12, 2012, http://www.epa.gov/cleanenergy/energy-and-you/affect/air-emissions.html.

selves are also recaptured and reused without being degraded to a lower-quality product. All manufacturing leftovers and final products are 100 percent environmentally benign. Such leftovers are directly reusable, ecologically neutral, or suitable for aerobic or anaerobic digestion by bacteria or other organisms, which would break it down into harmless (or, better yet, useful) materials.

At least, that is the goal. To attain it we at Greenway College will be counting on technological smarts, not on people's eagerness to be miserable. We respect those who are eager to radically simplify their lifestyle in order to achieve zero waste, and some studies even show that voluntary simplifiers may increase their happiness,[5] yet Greenway will not be predicated on anybody's willingness to give up basic comforts and conveniences. We envision a comfortable, spacious, and stimulating campus, generously appointed and lighted, where people work hard to design, test, document, and encourage environmentally benign and sustainable devices, from widgets to cities. This and other lofty environmental goals can, we believe, be achieved through progress in technology, without need for austerity.

Many of our professional graduates will, we hope, go on to work in industry and government, taking their expertise to ever-widening circles of society.

How Do We Do It?

Zero waste is achieved by meticulous attention to every aspect of consumption: buildings, manufactured objects,

---

5   Stephanie Rosenbloom, "But Will It Make You Happy?" *New York Times*, August 7, 2010, accessed May 23, 2012, http://www.nytimes.com/2010/08/08/business/08consume.html?pagewanted=all. See also Samuel Alexander and Simon Ussher, "The Voluntary Simplicity Movement: A Multi-National Survey Analysis in Theoretical Context," the Simplicity Institute, 2011, accessed May 23, 2012, http://simplicityinstitute.org/pub/The-Voluntary-Simplicity-Movement.pdf.

recycling, composting and wastewater, air pollution, hazardous waste, and, of course, energy in all its forms. Below, we glance at technologies for zero waste in each of these departments (energy generation and storage are covered in more detail in chapters 2 and 3). Finally, we outline of how these methods might be woven together at Greenway.

Buildings. Buildings and what goes on inside them—heating, cooling, cooking, computing, and the like—are major consumers of energy. Forty percent of all US primary energy use and 70 percent of all electricity use occurs within residential and commercial-class buildings. "Commercial"-class buildings include the nonresidential buildings one finds on campuses, such as libraries.[6] A large chunk of any university's budget thus goes for heating, cooling, and electricity consumed in buildings. Yet in no other area of energy consumption is there so much demonstrated room for improvement.[7]

The Passivhaus standard, invented in Germany, shows us how to build houses and other structures that need no large gas or oil burners, electric heaters, air conditioners, wood stoves, or other energy-eating technologies to remain fresh, well lighted, and at a comfortable temperature year-round. Even in the cool, cloudy climate of northern Europe (or the hot, sunny climate of southern Europe) these structures remain fresh, light, and warm. They achieve these goals largely through the use of passive solar heating, heat-exchanging air circulation, and insulation. The up-front cost of a

---

6   Paul Torcellini, Chad Lobato, and Tom Hootman, "Main Street Net-Zero Energy Buildings: The Zero Energy Method in Concept and Practice," NREL, May 2010, accessed December 1, 2011, http://www.nrel.gov/sustainable_nrel/pdfs/47870.pdf.

7   M. Levine et al., "Residential and Commercial Buildings," in *Climate Change 2007: Mitigation.* Contribution of Working Group III to the Fourth Assessment Report of the Intergovernmental Panel on Climate Change, ed. B. Metz, O. R. Davidson, P. R. Bosch, R. Dave, L. A. Meyer (Cambridge and New York: Cambridge University Press, 2007).

passive house's energy-saving features are largely recouped by minimizing expensive, unneeded heating and cooling systems. In Europe, tens of thousands of buildings already meet the Passivhaus standard; construction in the US lags in part because the high-efficiency doors and windows required are not widely available on the US market.[8]

In the US, several ways have been developed to rank buildings on their use of energy and other resources. The best known of these methods is probably the Leadership in Energy and Environmental Design (LEED) standard developed by the US Green Building Council, a nonprofit trade council. LEED ranks individual buildings on a number scale by awarding points for low-resource construction, water usage, energy efficiency, and other "green" features. At successively higher LEED point levels, a building can be deemed Certified, Silver, Gold, or Platinum. A LEED Platinum building approaches the net-zero ideal in energy and water usage, but Greenway College will strive to rethink and far exceed the LEED certification checklists.

The LEED certification levels are perhaps unfortunately named, since they may suggest that a Gold building is necessarily more expensive than a Silver, and a Platinum *very* expensive. Fortunately, this is not true: properly engineered efficiency is *less* expensive than inefficiency. The US National Renewable Energy Laboratory, putting its money where its mouth is, has erected six large, LEED-Platinum office buildings totaling over 300,000 square feet. Solar panels enable the facilities to use zero net grid electricity and zero on-site fuel for heating or cooling. NREL's per-square-foot construction costs were above average for commercial space, but the

---

8　　Tom Zeller, Jr., "Can We Build in a Brighter Shade of Green?" *New York Times*, September 25, 2010, accessed December 1, 2011, http://www.ny-times.com/2010/09/26/business/energy-environment/26smart.html?sq=passivhaus&st=cse&adxnnl=1&scp=1&adxnnlx=1322755609-tH7jZG7KFpp-GTHS1CDDNnA.

buildings' unusual features will have a simple payback time of around five years,[9] after which the facility will have been earning *net profit* on those features. Ithaca College's Park Center for Business and Sustainable Enterprise, completed in 2008, and also LEED Platinum, incurred an up-front construction premium of only 5 percent and quickly recouped that investment from low operating expenses.[10] "Green," in building, is also the color of money.

A skeptic may ask: If net-zero buildings are so economical, why isn't everyone building them? There are several reasons. One is the ubiquity of split incentives: for example, if you're building a structure but will not be its future tenant, you will not be paying the energy bills, so why sweat the efficiency details? In general, however, the biggest obstacles are lack of knowledge and cultural inertia. Most builders and building buyers simply do not know that net-zero buildings are possible, or assume that they would be super costly and inconvenient. Our society's assumptions lag its actual know-how. In addition, since inefficient designs dominate the US mass market, efficient designs often are custom designs and therefore *do* cost more in part due to customization. As efficient designs and materials become the norm, their cost tends to decrease. A classic example is the refrigerator, which has increased in efficiency fourfold since the 1970s while it has shrunk in cost by two-thirds (in 2010 dollars), grown in capacity, and acquired more features.[11]

---

9   "Laboratories for the 21st Century: Case Studies: NREL Science and Technology Facility, Golden, CO," EPA/DOE, undated, accessed December 1, 2011, http://www.nrel.gov/docs/fy10osti/47662.pdf.
10   Peter W. Bardaglio, "To LEED or Not to LEED," *Today's Campus*, 2011, accessed December 1, 2011, http://todayscampus.com/articles/load.aspx?art=1823.
11   "BTP Drives Minimum Refrigerator Energy Use Standards to Save Consumers Money," DOE EERE program website, accessed June 19, 2012, http://www1.eere.energy.gov/buildings/saving_energy_refrigerator.html.

Energy Use___ ___   Price •••   Volume ━━
(kWh/year)   (in 2009 $)   (cubic feet)

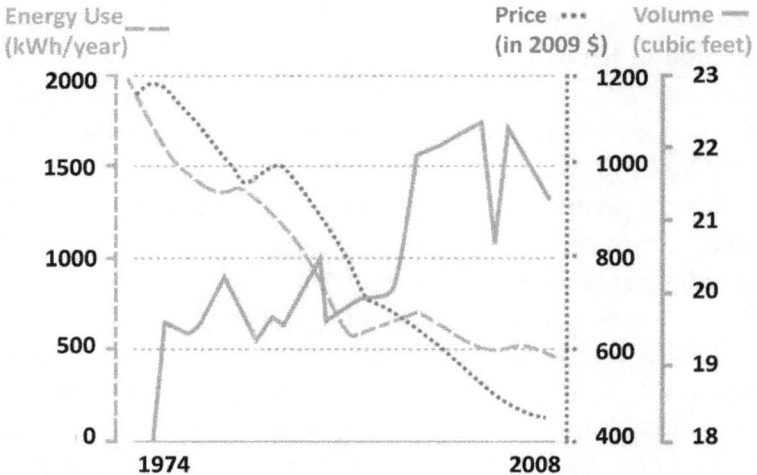

Fig. 4.1. Plot of the amazing success story of efficiency and the refrigerator. Long dashed line shows that average energy use for a new refrigerator in the USA has decreased fourfold since 1974; short dashed line shows that at the same time, average cost has decreased over two times; and solid line shows that both of these improvements in efficiency and cost occurred even while average refrigerator size grew substantially. (Source: US DOE EERE.)

In recent years, a number of colleges have seen how important sustainability is to incoming students and invested in green technology. The rapidly decreasing cost of solar energy especially has made this increasingly feasible. In 2018, American University became the first university in the United States to be certified as carbon neutral. The campus has six LEED-certified buildings, sources half of its energy from three solar arrays totaling 250,000 panels and half from renewable energy credits, and is renovating its heat and hot-water system.[12] Even this impressive facility still leaves room for improvement: it makes no claim to be zero *waste* (only zero net emissions, not solid or liquid waste) and is grid-con-

---

12 American University, "Carbon Neutrality Is Now Reality at American University," April 25, 2018, https://www.american.edu/media/news/20180425-carbon-neutrality.cfm.

nected. Yet it is important proof—one of many examples that could be cited—that the goals we are setting are not fantastic. Net-zero building technologies already exist: we need only apply them (and then spend the next decades figuring out how to make them even more elegant, effective, and profitable than they already are—chasing the refrigerator curves, as it were, of manyfold increased efficiency, lower costs, and improved features). Greenway will take advantage of lessons learned by early adopters like American and NREL, but will push the bar higher and yet higher, using newer technology, better designs, and smarter strategies.

When a site for Greenway has been selected and funds are in hand, we will hire an architectural firm with ample green-building experience to design a state-of-the-art green campus. Its structures will be traditional college buildings in outward appearance, but will *exceed* the standards of a 100 percent LEED Platinum campus (which does not yet exist anywhere). Leapfrogging LEED is our goal because LEED, with its point awards for Sustainable Site, Water Efficiency, Energy and Atmosphere, Materials and Resources, and Indoor Environmental Quality, encourages an *incremental* approach to sustainability. The LEED checklists are superb, but in building Greenway we have a chance to blow past them—indeed, to abandon checklists entirely for an all-at-once approach that will achieve better results for lower costs. Green aspects of Greenway's built-to-last buildings will include solar positioning and window sizing, super-insulation, computer-controlled air freshening and heat exchange, advanced windows, daylighting (which, besides saving money, improves student health and academic performance[13]), high-efficiency artifi-

---

13   "Daylighting in Schools: Improving Student Performance and Health at a Price Schools Can Afford," NREL/CP-550-28049, US National Renewable Energy Laboratory, August 2000, accessed March 15, 2012, http://www.nrel. gov/docs/fy00osti/28049.pdf.

cial lighting, open architecture for visualization of plumbing, mechanical, and electrical workings, and much more.

Greenway's concern for sustainability and zero waste will begin with construction itself. Contractors will be identified who are willing to enter into zero-waste agreements with Greenway, to include minimization and reuse of construction waste. For future study and reference, Greenway's construction will be thoroughly documented, preferably with the help of a film crew.

Manufactured Objects, Solid Waste, and Recycling

A college campus is in some ways a microcosm of the consumer side of industrial society. Therefore, in addressing the goal of zero waste at the level of the Greenway campus it behooves us to consider the waste experience of our society as a whole.

In the United States, we generate a huge amount of solid waste—400 million tons of industrial solid waste and 250 million tons of municipal solid waste. Only about 34 percent of US solid waste is recycled.[14]

Municipal solid waste, which is the sort that a typical college campus would be expected to produce, is a mix of every possible material—glass, plastics, organics, metals, paper, wood, concrete, chemicals, you name it. To reuse, optimally recycle, or compost its component parts, this mixed waste stream must be separated. Efficient sorting is, therefore, key to successful recycling of solid waste. The most economic options for efficient sorting are hand-sorting by individual consumers (as in some curbside recycling programs) and automated sorting, which is sometimes combined with hand-sorting at a centralized facility. A disadvantage of consumer sorting is

---

14   Municipal solid waste figures from "Municipal Solid Waste," US Environmental Protection Agency, accessed March 12, 2012, http://www.epa.gov/epawaste/nonhaz/municipal/index.htm.

that single errors can ruin a batch of product: just one chunk of the wrong plastic in a bin can render a whole load worthless. Bar-coding and built-in radio-frequency ID tags may someday enhance the purity of auto-sorted materials, as might manufacturer return of used products, since manufacturers have intimate knowledge of their products' constituents.

Properly done, washing and reuse (of bottles and dishes, for instance) always uses less energy than recycling or throwaway alternatives, and where direct reuse of objects is not possible, recycling of relatively pure, fusible materials like glass, steel, and aluminum is straightforward. Recycling of paper and plastics is also possible, though these typically result in lower-quality products than the original. Greenway will, of course, engage in these classic forms of reuse and recycling.

However, most manufactured objects are impossible to assign to a recycling bin. Computers and other electronic devices, for example, mingle glass, metals, and plastics in a nearly inextricable fabric. Various plastics are often mingled in single products in a manner that prevents each one from being truly recycled: mixed plastics are reused, if at all, as raw material of lower grade. This is not true "recycling" but, as William McDonough and Michael Braungart term it in their seminal *Cradle to Cradle* (2002), "downcycling." Downcycling merely postpones a material's trip to the landfill by one or two usage cycles. The goal of true *re*cycling should be to cycle materials back into materials of comparable or even higher value, indefinitely. This ideal is not always attainable: some materials, like glass and pure metals, are essentially immortal, but others, like paper fibers (which shorten with each trip through the paper-making process), can be reused only so many times for a single function. Yet, the lifecycle of paper may be adjusted so that no toxins are released and each fiber ends in the compost heap or biomass power plant after its life as paper.

A large part of the answer to the problem of impure,

mixed, and inextricable materials lies in product design. The designer should look at the whole "cradle to cradle" lifecycle, taking into account all costs from raw-material extraction to reuse, recycling, or final disposal. Unfortunately, in most industries, once a plastic toy, a French fry container, a bottle of motor oil, or an automobile leave its place of assembly, the original manufacturer has practically zero responsibility for it. Accordingly, current standard industrial practice is to look only at the cost of manufacturing and to try to meet environmental regulations at the manufacturing site. There is no incentive—beyond PR and sheer goodness of heart—for a company to exceed these regulations or to consider the whole product lifecycle.

Nevertheless, in recent years the cradle-to-cradle (C2C) concept has been translated into technical standards and is being taken with increasing seriousness by a number of manufacturers, including Nike and Ford.[15] Cisco, Apple, and other companies have instituted end-of-life, product-take-back programs that begin to close the product lifecycle loop, albeit imperfectly. It is a small beginning, but a beginning.

The task at hand touches on the whole globalized system of resource extraction, product manufacture, and material disposal or reuse—a challenge that Greenway will exist, in part, to help address. Clearly, Greenway cannot singlehandedly reform the manufacturing habits of the planet, and so at least in the beginning will be obliged to participate to some extent in downcycling and the waste production embedded in the making of many essential complex items for an educational setting such as computers, maintenance equipment, and even light bulbs. However, Greenway is intended to be a living laboratory, not a museum piece: it will reflect the

---

15   Reena Jana, "New Clout for Cradle to Cradle Design," *Bloomberg Businessweek*, September 19, 2007, accessed December 2, 2011, http://www.businessweek.com/innovate/content/sep2007/id20070919_689774.htm.

state of the art, not the stasis of the art. We will therefore welcome engagement with thorny issues such as downcycling and e-waste, rather than shunning them.

Certain positive measures can be already be foreseen. For example, we will strongly favor suppliers of chemicals and objects that can engage in C2C-type takeback programs. In some cases we may be able to negotiate bulk-purchase agreements conditioned on takeback and recycling. We will not only continually monitor, analyze, and reinvent our own usage patterns, but will also encourage and empower profitable green engineering by developing new technologies, cataloging technologies and their costs, providing education and tools for good analysis, and presenting awards to companies that demonstrate new technologies. The keynotes will be tracking of materials, lifecycle analysis of products, and constructive engagement with manufacturers and suppliers.

As an institution, Greenway College will not participate in legislative lobbying or political activism (though our individual community members will, of course, do as they please). Rather, Greenway will work strictly within the parameters of technology and markets. We believe that good design, uncoerced social conscience, and market forces will join to make our demonstrations and quest a successful one.

Organic Wastes and Wastewater

Graywater (mildly contaminated water), blackwater (sewage), kitchen wastes, and other organic materials such as lawn clippings and agricultural leftovers have one thing in common: bacteria love them. Our waste is their feast. By various methods, including compost heaps and anaerobic (no-oxygen) digester tanks, such wastes can be made harmless and useful at the same time: waste becomes a resource.

When bacterial digestion takes place in the absence of oxygen, as it typically does in liquids, the major gaseous product is methane (CH_4), a useful fuel. Biodigestion is widely used across Europe, and New York City already harvests half the methane released by its vast wastewater-treatment system and burns it to generate about $10 million of electricity per year.[16] Aerobic composting allows oxygen-breathing (aerobic) bacteria to do a similar job and produces carbon dioxide rather than methane. (Wetter composts, or large piles, can produce some methane.) In general, liquids containing organic waste are *anaerobically* digested, reducing waste to methane, while solid wastes are composted *aerobically*, producing generally carbon dioxide. In particular, aerobic composting produces temperatures on the order of 70° C (160° F), which kill pathogens (bacteria, viruses, fungi, parasites) and plant seeds from organic wastes (including human excrement and food wastes). Both kinds of composting transform unpleasant, troublesome wastes into reduced volumes of potentially useful solid residue.

The solid residue from a properly designed anaerobic digester or compost system has no offensive odor, contains fewer harmful bacteria than its inputs (or none), and, depending on source material and process, can be used as fertilizer, a soil amendment, or animal bedding. Recology, a company that composts 10,000 metric tons of material for San Francisco each year, sells most of its rich, black product at a premium to the California wine industry.[17] Your last glass of California merlot probably consisted, in part, of recycled restaurant scraps . . . proof that the zero-waste economy will not be without its pleasures.

In theory, 55 to 70 percent of the weight of a typical

---

16   Mireya Navarro, "City Is Looking at Sewage Treatment as a Source of Energy," *New York Times*, February 8, 2011, accessed December 2, 2011, http://www.nytimes.com/2011/02/09/science/09sewage.html.
17   Jim Carlton, "San Francisco Garbage Helps Make Vineyards Thrive," *Wall Street Journal*, October 31, 2011, accessed December 2, 2011, http://online.wsj.com/article/SB100014240529702036331045766216332426080 82.html.

municipal waste stream—which Greenway's will resemble—can be composted.[18] Not all of it may be ideal, however, as it may contain impurities. As with various other forms of recycling and zero-wasting, separation earlier in the stream and decreasing the mingling of disparate materials increases recovered-material purity and therefore value while decreasing downstream processing costs.

Greenway will employ a mix of well-tested technologies to recycle valuable organic materials. Humus from composting will be sold to gardeners or farmers or else used for on-campus farming and landscaping. An anaerobic biodigester will process blackwater (sewage) into useful methane and harmless compost, and can also be fed with cellulosic materials such as wood harvested sustainably on-site and paper. Methane can be either burned directly for heat and power (releasing CO_2 in the process) or "reformed" to produce CO_2 and hydrogen, where the latter can be used as an ultraclean fuel in fuel cells and engines. Greenway may, in fact, choose to put *all* its organic solids into biodigestion, to squeeze the maximum fuel and compost value out of every ounce; but such decisions will depend on engineering analysis during preconstruction campus design.

Wastewater processing is most effective when graywater is pre-separated from blackwater, because graywater is about 50 to 80 percent of residential wastewater, has a much lower nitrogen content, can be stabilized faster, and contains far fewer pathogens. Graywater treatment options include landscape irrigation, greenhouse infiltration beds, or constructed wetlands. These methods process the graywater in a shorter time than needed for effective blackwater treatment and can also be visually attractive.

---

18   Mitch Renkow and A. Robert Rubin, "Does Municipal Solid Waste Composting Make Economic Sense?" in *Waste Management and Planning*, ed. R. Kerry Turner, Ian Bateman, and Jane Powell (Cheltenham, UK: Edward Elgar Publishing Ltd., 2001), 15–38.

Greenway will find a high-grade use for every ounce of organic matter produced on campus, from potato peels to poop. Every gallon of water Greenway releases to its environment will be at least as clean as when it was pumped in.

Hazardous Waste

Waste is deemed "hazardous" if it is flammable, reactive, corrosive, toxic, or radioactive. Our ultimate goal is to produce zero hazardous waste not only on-site, but throughout the supply chains of all foods and manufactured products consumed on campus. A realistic first goal, however, will be zero on-site hazardous waste production coupled with *minimization* of hazardous-waste production by upstream suppliers.

Most hazardous "wastes" are actually valuable resources, and their disposal can be expensive. It is therefore often in the financial interest of an industry to recapture and consume all such materials. Even apart from the financial incentives for reducing hazardous waste, in many instances it is only a matter of proper engineering and design to manufacture products while producing zero hazardous waste. Greenway will push for cleaner production while helping manufacturers understand its technology.

Air Pollution

Most air pollution consists of the six "criteria" pollutants that the Clean Air Act requires the Environmental Protection Agency to regulate: sulfur dioxide (SO_2), nitrous oxides (NO_x), carbon monoxide (CO), ozone (O_3), lead, and particulates (soot, smoke). Air pollution drains between $71 billion and $277 billion from the US economy per year, over 95 percent of it in health costs.[19]

---

19   "Pollution: Costs of Inaction," *OECD Observer* 263, October 2007, accessed December 1, 2011, http://www.oecdobserver.org/news/fullstory.php/aid/2351/.

We will produce zero air pollution on Greenway College campus and will work at first to minimize—ultimately, to zero out—air pollution in the manufacture of the products we purchase. We will record emissions produced during manufacture of products used at Greenway and maintain an up-to-date, open-access record of these data on the college website. The list will be broken down into lifecycle emissions from purchased products and any on-site emissions. Indeed, web pages will be maintained for all waste products, energy consumption and production, and other inputs and outputs of the campus system. These dashboard-type accounting tools will make Greenway's successes and challenges transparent to the community, encouraging and informing efforts to improve.

Carbon dioxide, CO_2, falls into a special category. Until a few decades ago, CO_2 was not considered a pollutant at all, since it is a natural part of the Earth's atmosphere. However, even clean, fresh water must be considered a problem—call it a "pollutant" or whatever you like—if you are drowning in it. Drastic increases in human-caused CO_2 output since the mid-nineteenth century, mostly from fossil fuels, have increased CO_2 in the atmosphere by almost 40 percent over its preindustrial value, to a level not seen in fifteen million years.[20] In recent years the rate of increase of global emissions has accelerated, and there is no end in sight.[21] Some activities at Greenway will inevitably release some CO_2, but others will utilize CO_2, such as plant life, including an appropriately sized on-campus woodland, which can be sustain-

---

20   "Last Time Carbon Dioxide Levels Were This High: 15 Million Years Ago, Scientists Report," *ScienceDaily*, October 8, 2009, accessed December 2, 2011, http://www.sciencedaily.com/releases/2009/10/091008152242.htm.

21   "CO_2 Increase in Atmosphere Accelerating," Associated Press, November 21, 2011, accessed March 15, 2012, http://www.cbc.ca/news/technology/story/2011/11/21/environment-wmo-greenhouse-gas.html.

ably managed for forest products, recreation, wilderness, and wildlife—allowing for on-site net zero CO_2 emissions.[22]

The largest US source of air pollution (all types included, not just CO_2) is energy generation—burning of coal for electricity and gasoline for transportation. However, Greenway's zero grid-electricity plan (see chapter 2) and energy storage plans (see chapter 3) will ensure the production of zero electricity-related and transportation-related emissions including zero net CO_2 emissions.

The Never-Ending Quest

Zero output of waste—gaseous, liquid, and solid, whether on campus or in the network of supply—will be a continuing challenge to attain. Indeed, it is a part of the zero-waste program to always be striving toward complete elimination of waste and moving toward maximal resource utilization. Our journey toward zero waste will be just beginning when Greenway opens its doors. The zero-waste mission will never be absolutely achieved because opportunities will multiply, rather than be used up, as technology progresses. A large part of Greenway's mission will be to apply, develop, and learn from zero-waste processes, technologies, and design principles through many decades to come.

---

22 Simultaneous management for all these values is feasible: see R.W. Malmsheimer et al., "Managing Forests Because Carbon Matters: Integrating Energy, Products, and Land Management Policy," *Journal of Forestry*, October–November 2011, accessed December 2, 2011, http://www.safnet.org/documents/JOFSupplement.pdf.

Castles on the Ground

The future is not completely beyond our control. It is the work of our own hands.

—Robert F. Kennedy

If you have built castles in the air, your work need not be lost; that is where they should be. Now put the foundations under them.

—Henry Thoreau, *Walden* (1854)

Greenway College will be a totally "green," zero-sacrifice learning community that runs entirely on clean, renewable energy, independent of the grid. It will implement zero-waste principles throughout its fabric while providing all the services that one would expect from any top-drawer four-year college, and will do so as reliably, conveniently, and affordably as any grid-connected, waste-producing campus. Although a number of net-zero or energy-independent housing blocks and individual buildings have been built to date—all important early adoptions and test-beds—Greenway will be the first *zero-waste, energy-independent, 100 percent renewably powered college in the world.* It will also be the first college dedicated to the study, advancement, and spread of such technologies.

Any college, no matter how physically green or clean, is nothing without faculty and students. For faculty, Greenway will gather top experts and teachers in alternative energy generation, energy storage, zero-waste systems, and other areas of sustainable engineering. We anticipate that our unique laboratory-campus and ambitious mission will enable us to attract top talent in these fields. Likewise, we expect that Greenway will have no trouble filling its roster with qualified students eager to learn in a stimulating, interactive, directed-study, apprenticeship-based style.

In previous chapters we've shown that in the broad areas of energy independence and zero waste, the technologies needed to build Greenway already exist. Although a full-blown, in-detail campus design will require input from architects, engineers, accountants, faculty, and other experts and stakeholders, including members of Greenway's proposed home community, it is possible to outline a workable vision of Greenway campus right now. The final blueprints will differ, but in detail rather than spirit. So let's put meat on the bones. What will Greenway College actually look like? How will it work?

Land

If faculty and students are the lifeblood of any school, grounds and buildings are its bones. Most fundamental of all is the land on which everything stands. Greenway College could be built on a wide range of landscapes, since most of its green technologies do not depend on special geographic features, but this does not mean that Greenway or any similar enterprise ignores its home place: on the contrary, Greenway's design will be fitted to work with its particular site, making it green literally from the ground up. An example of such a site-friendly design approach has been set by the largest zero-net-energy office building in the United States, the National Renewable Energy Laboratory's 360,000-square-foot Research Support Facility in Golden, Colorado (completed 2010). The new facility was designed to fit into the natural character of the mesa site, minimally disturb the terrain, and conserve water resources.[1]

Greenway could be built on undeveloped land or on a predeveloped or partially developed site. It might retrofit old structures or scrape them off and build from scratch, as needed. The possibilities are endless, but to make this discussion specific, we envision purchase of a closed-down industrial, medical, recreational, or other campus that is adjacent to or includes a fairly large tract of undeveloped land. An ideal site would be in the range of 200 to 1,000 acres, enough to allow for future expansion with permanent preservation of open and forested space. The actual college will occupy a fraction of this land, starting with about 50 acres and eventually expanding to about 100. Purchase of an obsolete, crumbling facility would tend to hold land cost down, preferably under $10 million, and assure that real estate is not a dominant share of overall cost.

For specificity, we can imagine building Greenway on a closed-down college in southwestern Vermont—an actual property that listed a few years ago for under $10 million. (Greenway could be built in any climate and setting within the country.) This particular 400-acre site has a number of desirable features, including existing college facilities, good solar exposure, a combination of land at low elevation and acreage on a moderately elevated ridge, the latter appropriate for both wind turbines and pumped hydroelectric energy storage. Pumped hydro is the lowest-cost, longest-lifetime, most-tested technology for large-scale energy storage presently available, but where it is not feasible, other technologies can be used, including compressed air, hydrogen, thermal, and large batteries.

---

1   "The Design-Build Process for the Research Support Facility," US National Renewable Energy Laboratory, https://www.nrel.gov/docs/fy12osti/51387.pdf.

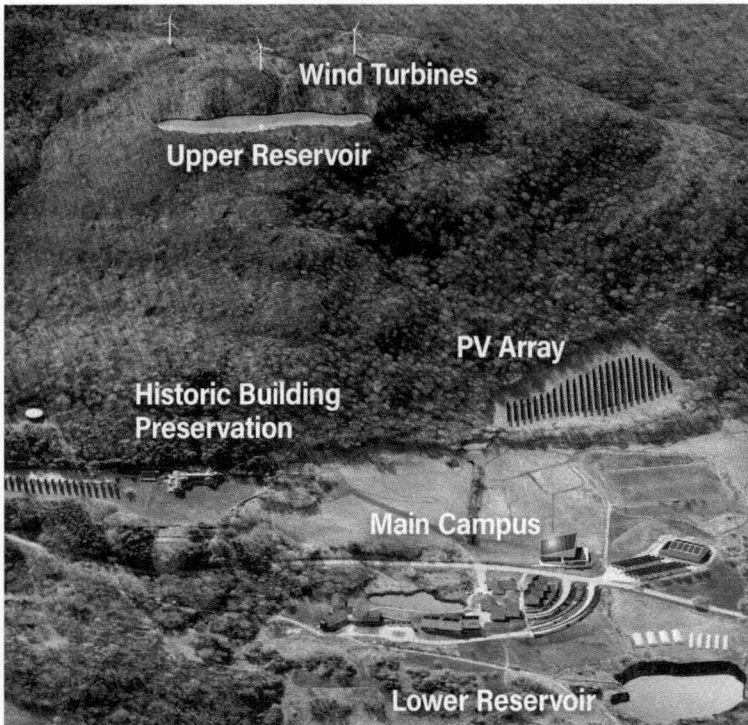

Fig. 5.1. Possible Greenway College layout based on the actual topology of a closed-down, 400-acre college in southwestern Vermont. Wind turbines and buildings are exaggerated in size for visibility.

A high-quality wind resource and siting for pumped-hydro energy storage are not necessary for building a stand-alone, totally green campus, but are offered by many plausible sites and are assumed here. And, although the water and wind resources of this site may make it easier to envisage meeting our clean energy goals, the climate compensates: Vermont has the sixth-coldest winters of any US state.[2] A New Mexico

---

2   Liz Osborn, "Coldest States in America," Current Results, accessed January 5, 2012, http://www.currentresults.com/Weather-Extremes/US/coldest-states.php.

Greenway campus would have a lot less water to play with, but a lot more solar energy, and (depending on location) perhaps just as much wind. Figure 5.1 shows a possible Vermont Greenway layout based on our exemplary 400-acre site.

The open and wooded lands of the campus wooded portion of the campus will provide fields, gardens, watershed, recreation, habitat, and possibly forestry products such as wood for power generation and liquid biofuel manufacture. A sustainable forestry program could provide many benefits to the campus while also contributing to the local economy.

Buildings

We do not propose upgrading and constructing the best buildings *possible*. Building ideals shift year by year as our know-how grows: the best possible building of today differs in many ways from the best possible building of 1990. Besides, if we already knew how to build Greenway perfectly there would be no point in building it, because we would have nothing to learn. Yet we know enough to begin: we can build a zero-waste, energy-independent, nonideal campus without busting the bank. Our physical plant's ongoing imperfection or non-finality will be a feature, not a bug, for Greenway is intended be a laboratory, a hands-on think tank, a learning and teaching community, a *college*, not a place where students plug their heads in to receive a download of finalized information.

If Greenway is to be a place of learning, it must be built accordingly. While its buildings will incorporate enduring principles of sustainable architecture (self-heating, self-cooling, best siting, optimal solar orientation, and the like), on top of that timeless armature of good design features they will feature as many as possible of the best cutting-edge technologies for generating and using power and light, for handling clean and dirty water, and the like.

Many historical colleges can be traced back to a first building or small set of buildings. These often served as the entire college—dining halls, dormitories, classrooms, laboratories, offices. Greenway will begin this way, too: compact but complete.

The core or original campus will consist of upgrading existing buildings to state-of-the-art, efficient clean-energy-powered buildings and construction of one new building. Existing buildings that will be upgraded include: (1) gymnasium, (2) residence buildings, (3) combined classrooms, residence, and student center, (4) cafeteria, and (5) combined offices, classrooms, and laboratories (see figure 5.2). One new-construction main building will be added with public education facilities, auditorium, and some state-of-the-art teaching-project spaces and laboratories. The core buildings will have about 28,000 square meters (m²)—about 300,000 ft²—of floor space and will be upgraded for best performance of their energy, lighting, and other features, such as their windows and rooftop solar panels.

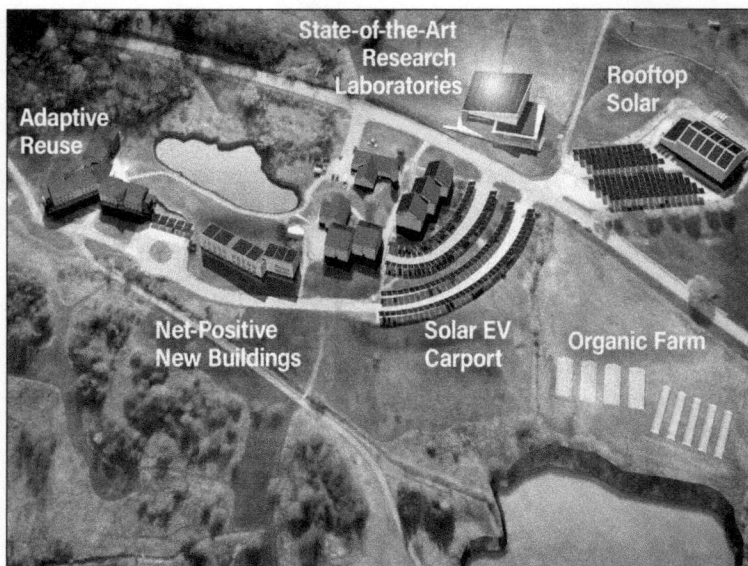

Fig. 5.2. Proposed Greenway College core campus. buildings.

The residence hall will house approximately 180 students, four faculty or staff families, and two caretakers. Remaining faculty and staff will find housing off-campus, as is common at colleges and universities; this will link us to the surrounding community and benefit its economy. There will be about fourteen classrooms, one auditorium, twenty-five offices, and ten teaching laboratories. The teaching laboratories will contain standard equipment in physics, chemistry, electronics, and other disciplines, but in addition will be connected, either directly or electronically, to the working mechanisms and life-metrics of the college: water flows and volumes; brightness and shadow levels; every form of energy generation, usage, and storage; inside and outside temperatures, winds, and light levels; structural loads. All will be treated as functional experiments, subjects for study. The physical fabric of Greenway's campus will, like a spacecraft, be permeated by a veritable nervous system of sensors and links that enable its users to track, learn from, and ultimately improve every aspect of its functioning. The campus will obey the Greek dictum to "know thyself" as few institutions have ever done—and thanks to modern computers and wireless sensors, it will do so at modest cost.

Facilities will include a dining hall and snack bar, student fitness and lounge area, bookstore, public information and display center, administrative offices, restrooms, and possibly a rooftop observatory. All working equipment will be accessible by the faculty for demonstration, examination, and tours. The goal, at Greenway as at any modern university, is simple: comfortable, well-lighted, convenient spaces in which to work, play, live, and learn. Construction costs for the buildings, exclusive of energy generation and other sustainable technology, will be around $50 million.

During and following the upgrade and construction of these first buildings, as well as all of future modifications and expansions, students and faculty will be directly involved in studying them from the standpoints of science, engineering, economics, ergonomics, and all other relevant disciplines. In this way, their lived experience will be a part of their studies of green technologies and engineering.

During and after construction, use of hazardous materials will be minimized. In routine operation, consumer garbage will be eliminated. Reusable or compostable containers will be used for all packaging. Shipments from outside companies will be attached to agreements for return of unwanted excess shipping containers and recovery of obsolete products. All outside purchases will be analyzed for environmental impact during fabrication, use, and recovery/recycling/reuse after their intended lifecycle. All excess food products, as well as other safe and rapidly biodegradable products, including human excrement, will be unobtrusively composted or digested. All excess materials will be completely dealt with on-site or returned to the manufacturer to eliminate waste.

Energy and Waste: Overview

Energy brings buildings to life: it lights, warms, cools, circulates air, and runs computers, phones, elevators, water pumps, and other devices. One form of energy—electricity—can perform all these jobs, but not always economically. Electrical energy is the most refined and most versatile of energy forms—but not necessarily the least expensive or most direct way to perform a function. We will use it—with maximum affordable efficiency—at every point where its use is appropriate, and avoid it wherever there is a more direct, less costly alternative. The result is that Greenway's internal energy economy will be a flexible tapestry adapted to the full range of end uses. Electricity will run computers, communications equipment, artificial light sources, pumps and fans, and other devices that require it. The sun will provide most heating and lighting and virtually all hot water. Insulation, thermal masses, sun shading and reflection, and air circulation will provide all or nearly all of needed space cooling (as these techniques already do at NREL's LEED Platinum buildings in Golden, Colorado, as well as in Europe's tens of thousands of Passivhaus structures). If necessary, additional ground-loop heat pumps (discussed further below) will provide extra heating and cooling capacity to assure that Greenway's buildings remain comfortable at all times, without fail. Gaseous and liquid fuels generated on-site, along with electricity, will be used to run on-campus vehicles, cooking equipment, refrigerators, and other miscellaneous devices.

Electricity will be generated mostly by solar panels and, depending on permitting, a wind-turbine installation, with lesser amounts of additional forms of generation potentially also installed (for example, a biomass-burning system providing combined heat and power, or fuel cells running on hydrogen). Bulk energy storage may be primarily in a two-acre water reservoir located near the wind-turbine installation. A wastewater-treatment greenhouse with artificial wetland and compost facilities—built for strict odor control, as are all good compost sites—will be showcased not far from the main buildings. All facilities will be viewable by tour and designed for maximal accessibility. Gasoline-powered vehicles will be parked at the campus periphery, and a small fleet of electric, hydrogen-combustion and fuel-cell vehicles, charged from on-site power sources, will be available for on-campus transport of materials and mobility-impaired people.

No corners will be cut when it comes to reliability of any campus systems; on the contrary, extra effort and expense will be undertaken to foresee and avoid, or blunt the impact of, possible system failures. All operations will be to some extent overengineered, with backups to compensate for inevitable failures. For example, a hydrogen-burning backup generator will supply power if all else fails. (There will be no grid connection, except possibly a remote laboratory for experimental purposes.) This extra up-front investment will not only serve the Greenway community directly, but ensure the college's long-term success by preventing embarrassing lapses in function. It is up to us, in the design stage, to assure that the headline "Students at Sustainable College Shiver in Dark During Power Outage" *never* sees print.

Next, we review the basic physical systems mentioned above—energy supply and storage and zero-waste design—in more concrete detail.

Solar Power

Solar power, now approaching the least expensive electricity source, will be the leading means at Greenway of generating electricity. Solar could in fact be economically used to generate all of Greenway's power, but for redundancy, experiment, and instruction, we will install multiple technologies, including wind turbines—which balance intermittency to an extent and may reduce energy storage requirements. Solar equipment is very reliable, panels can be stationed atop already-developed land (rooftops, parking lots, and the like), and it raises no concerns about turbine noise, ridgeline development, or other controversies.

Photovoltaic panels turn sunlight directly into electricity. They have no moving parts, produce no noise or pollution during operation, are highly reliable, require little maintenance, and last thirty years or more. They generate the most energy when pointed directly at the sun at all times, but continuous, direct sun pointing is only possible for panels mounted on moving mechanisms, which cost more. In Vermont, tilting a panel at about thirty degrees and pointing it due south produces the most electricity annually for a stationary panel; sun-following trackers add mechanical complications but produce additional annual generation. As shown in figure 5.2, for Greenway we propose solar on most buildings and additional solar over parking lots and in some open spaces. Snow coverage of panels can be a concern in winter in northern latitudes, so ground-based panels will be installed either at a substantial angle or include tracking or seasonal adjustability.

Figure 5.3. At left, a wind farm in Searsburg, Vermont, that generates about 36,000 kWh/day—about five times the wind-power supply we envision for Greenway. Right, the National Renewable Energy Laboratory's Research Support Facility in Golden, Colorado. The facility's dark-blue rooftop solar panels generate about 1,800 kWh/day,[3] making the generation of this 2,800-m² system about 25 percent of the size proposed for Greenway.

---

3 Photo1: http://www.epa.gov/region1/eco/energy/re_wind.html. "Energy and Global Climate Change in New England," EPA, accessed Sept 2012. Searsburg site is approximately 6 MW with approximately a 25 percent capacity factor. Photo2: http://www.nrel.gov/data/pix/Jpegs/19089.jpg, accessed May 25, 2012.

Size and actual kilowatt-hour output of NREL system for all of 2011 used as basis of numbers given here: "System Advisor Model (SAM) Case Study: NREL Research Support Facility (RSF)," US National Renewable Energy Laboratory, 2012, accessed May 25, 2012, https://sam.nrel.gov/files/content/documents/pdf/NREL%20RSF%20Building%20Rooftop%20PV.pdf.

Capacity of solar system 1.6 MW: Heather Lammers, "Solar System Tops Off Efficient NREL Building," US National Renewable Energy Laboratory, September 29, 2010, accessed May 25, 2012, http://www.nrel.gov/news/features/feature_detail.cfm?feature_id=1516.

15.5% photovoltaic capacity factor for Colorado: R. Muren and C. Kutscher. "Analysis of Renewable Energy Deployment in Colorado by 2030," NREL Technical Report NREL/TP-550-42577, US National Renewable Energy Laboratory, December 2007, http://www.nrel.gov/docs/fy08osti/42577.pdf, accessed May 25, 2012.

For Greenway College, we propose a photovoltaic array rated for a maximum output of 1,000 to 2,000 kW. With a 15 percent capacity factor, such a 2,000-kW system would produce an average of 7,200 kWh/day—though all during the daytime and substantially more on sunny days during the summer. This entails around 14,000 m^2 (150,000 ft^2) of panels, an amount that can easily be accommodated on the site on rooftops, parking lots, and dedicated solar areas. At 2020 prices, such a system would cost about $6 million.

Wind

Wind power is highly dependent on wind speed: generation goes up roughly with the cube (third power) of wind speed, so doubling wind speed increases power output eightfold. In many areas, such as the western and central US, excellent wind conditions can be found in flat, wide-open spaces. Wind speed also increases with rotor height above the ground, which is one reason many modern wind turbines are so big. (Some large units have turbine hubs 140 meters—or 450 feet—above the ground, with blades that reach another 60 meters—or 200 feet—or more, making them significantly taller than the 170-meter Washington Monument). In other areas, including the northeastern US, the best wind conditions are found offshore and on ridgetops. Our example site for Greenway includes a ridge that can do double duty for wind turbines and hydroelectric energy storage. Ridgetop installations can have relatively small footprints (primarily for access roads and tower footings), but large turbine installations are often visible for many miles. Therefore, to reduce the visual impact, we propose installing not 200-meter turbines but more manageable midsize turbines rated at 400 to 800 kW with hub heights of at most 45 meters (150 feet) and blades rising another 24 meters (75 feet).

We visualize a nameplate 1,000 to 2,000 kW wind turbine setup with an expected 20 percent capacity factor.

Wind generators typically operate at a capacity factor (actual energy output as a fraction of maximum possible output) of 20 percent to 40 percent, depending on wind conditions, so a minimally well-sited 1,500 kW wind setup would generate, on average, approximately 7,200 kWh/day (sometimes more, sometimes less). Wind turbine costs are currently estimated at $1 to $3 per nameplate watt, installed, for systems less than 5 MW in capacity, making Greenway's expected wind cost somewhere around $4 million.

Combining a wind installation of 1,500 kW capacity and a photovoltaic system with 2,000 kW capacity will generate an around-the-clock average power of about 14,400 kWh/day, enough to run a fully-equipped, zero-sacrifice, energy-smart college campus.

Heat Pump

As noted above, air- and ground-source heat pumps may be used for heating and cooling. Ground-source systems, of which almost three million had been installed worldwide by 2010,[4] exploit the fact that, once you are a meter deep (3 feet) or so, the temperature of the next several hundred meters or so of the Earth is at an almost constant temperature year-round, about the average temperature for that region (e.g. 10–20 °C [50–70 °F] in most of the continental United States). This makes the Earth itself a potential source of heat in the winter and cooling in the summer, if leveraged by a relatively small amount of electric power (which drives the "pump" parts of a heat pump). Although sometimes called "geothermal" heat pumps, ground-source heat pumps have nothing to do with volcanic heat or hot springs; they work most anywhere.

---

4   Robert Crowe, "Demand for Geothermal Heat Pumps to Grow 14 percent by 2015," *Renewable Energy World*, January 14, 2011, accessed January 6, 2012, http://www.renewableenergyworld.com/rea/news/article/2011/01/geothermal-heat-pumps-demand-to-grow-14.

The primary space-heating installation will, then, be a ground-source heat pump that recirculates its own water ground-source loop (a "closed loop" system). Such a setup produces about 3 kilowatts of thermal energy for each kilowatt of electrical energy required to run the system. This does not break the laws of physics (which say that energy can only be transformed, not created or destroyed) because the extra thermal energy is extracted from the Earth, whose stock of thermal energy is large and replenished by the sun so that our heat-pump system cannot significantly diminish it.

The approximately 30,000 m^2 of inside space at Greenway will need a 200 kW system, assuming a 7 W/m^2 heating and cooling requirement. This might seem fantastically low to a conventional architect, since it is less than one-tenth the average for *total* energy use in a "commercial"-class space like an office building or hospital,[5] but is quite reasonable for at Greenway. The NREL Research Support Facility in Colorado, in actual operating experience, uses only 5 W/m^2 for heating and a tenth that for cooling.[6] The ground source heat pump system will also provide hot water, supplementing a solar hot water system.

Ground-source heat pump systems are estimated to cost around $1,000 per kW of capacity depending on size;[7] the price of Greenway's system is estimated at about $250,000. Four or five computer-controlled heat-pump units will be

5   "What is EUI?" US Environmental Protection Agency, accessed January 9, 2012, http://www.energystar.gov/index.cfm?fuseaction=buildingcontest.eui.

6   "NREL's Research Support Facility: An Energy Performance Update," US National Energy Research Laboratory, December 2011 accessed May 24, 2012, http://www.nrel.gov/sustainable_nrel/pdfs/rsf_operations.pdf. This document gives 8.58 kBtu/ft²/year for heating, and 8.58 kBtu/ft²/year divided by 8,769 hours per year, which gives .978 Btu/ft²/hr.

7   "Selecting and Installing a Geothermal Heat Pump System," US Department of Energy, 2011, accessed May 24, 2012, http://www.eere.energy.gov/consumer/your_home/space_heating_cooling/index.cfm/mytopic=12670.

installed, each of about 65 kW capacity, allowing for staggering of the heating load and maintaining a lower peak power. Such a system will cover less than an acre, and can go under a small field or parking area.

Ground Source Heat Pump Vertical Well

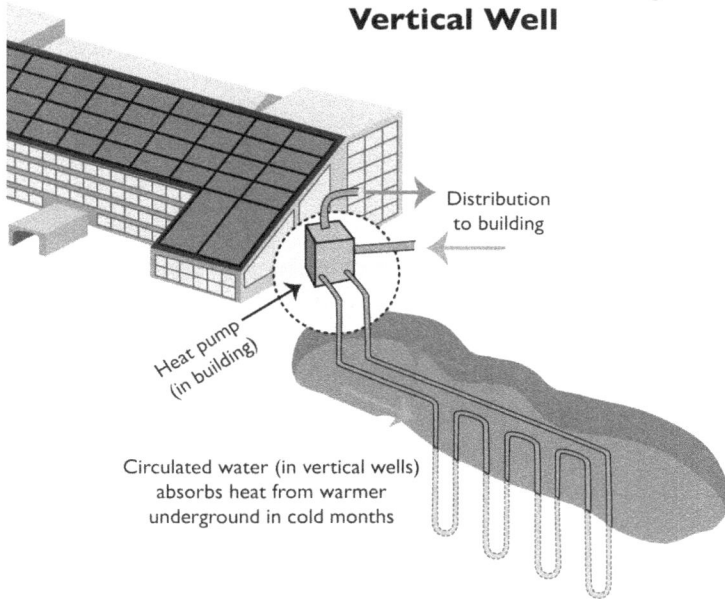

Distribution to building

Heat pump (in building)

Circulated water (in vertical wells) absorbs heat from warmer underground in cold months

Fig. 5.4. Illustration of a closed-loop vertical-pipe setup for a ground-source heating system.[8]

Direct Solar Heat

Passive solar design is the design of buildings to maximize the use of solar warmth in winter, ventilated cooling in summer, and solar lighting ("daylighting") year round. Overhangs will be constructed above most windows to block

8   "Geothermal Heat Pumps" at Energy.gov, accessed Oct, 2012, http://www.energysavers.gov/your_home/space_heating_cooling/index.cfm/mytopic=12650.

summer sun and allow autumn, winter, and spring sun to pass through windows facing every way but north. Advanced windows and automated systems for closing insulated curtains can decrease thermal losses at night. Light shelves can bounce daylight up to light-colored ceilings and thence down into workspaces, getting light where it needs to be while eliminating glare. Automatic sensors will dim or brighten overhead fixtures as needed to keep light levels even. Studies show that daylighting not only costs less than electric lighting, but also tends to improve worker mood and productivity.[9] Thermal masses—heavy chunks of the building, perhaps located at its lowest level—will store warmth for cool hours and coolness for hot hours. Costs for all these systems are included in the building budget; notably, although such buildings are marginally more expensive to build, they cost so much less to operate that the original investment pays back tenfold over building lifetime.[10] Money is not only invested but saved up front, as large conventional heating systems need not be purchased.

A moderate-sized active solar collector installation will be used to make hot water in the summer and to preheat water in the winter months. Millions of such systems are in use around the world. The installed cost of a 300 m² (3,230 ft²) solar hot-water system at Greenway will be about $200,000.

9   "Daylighting Resources—Productivity," Lighting Research Center, accessed May 24, 2012, http://www.lrc.rpi.edu/programs/daylighting/dr_productivity.asp.

10   Gregory Kats et al., "The Costs and Financial Benefits of Green Buildings: A Report to California's Sustainable Building Task Force," report developed for the California Sustainable Building Task Force, October 2003, accessed January 9, 2012, http://www.ciwmb.ca.gov/GreenBuilding/Design/CostBenefit/Report.pdf.

Bioenergy

Today, crude and unhealthy biomass-burning is still a major source of energy for hundreds of millions of poor people around the world, but cleaner, higher-tech ways of generating electricity and heat from crops, forestry products, and organic "wastes" are gaining popularity in industrialized countries. Net carbon dioxide (CO_2) emissions can be nearly zero for such fuels, as approximately equal quantities of CO_2 are absorbed by vegetation during growth as are emitted during combustion. In some cases, such as corn-based ethanol, there has been fierce debate over how much one really gains by making the biofuel, since fossil fuel is used to grow and fertilize the corn: according to the US Department of Agriculture, about 0.5 units of fossil-fuel energy must be invested to produce one unit of corn-ethanol energy. The CO_2 emissions of a gallon of standard corn ethanol are thus not zero, but about half those of a quantity of gasoline yielding the same energy.[11] Other issues with some specialized energy crops, including corn, are soil erosion, fertilizer pollution, and diversion of croplands from food to fuel production, with resulting higher food prices for the world's hungriest people.[12]

Fortunately, some biofuels can be designed and harvested to sidestep these problems. Greenway will implement only fully renewable, sustainable, local biofuel options.

During power generation with direct burning of biomass,

---

11   "2008 Energy Balance for the Corn Ethanol Industry," US Department of Agriculture, June 2010, accessed January 9, 2012, http://www.usda.gov/oce/reports/energy/2008Ethanol_June_final.pdf.

12   Environmental Working Group, "The Unintended Environmental Impacts of the Current Renewable Fuel Standard (RFS): A Guide to Common Sense RFS Policy," Fall 2007, accessed January 9, 2012, http://www.ewg.org/files/EWG_Corn_RFS_Fall_07.pdf. Also, "Ethanol Blamed for Record Food Prices," *Technology Review*, March 2011, accessed January 9, 2012, http://www.technologyreview.com/energy/37019/.

CO_2 and some pollutants are inevitably emitted—nitrous oxides (NO_x) and in some cases particulates (smoke and soot)—but overall lifecycle emissions from a new biomass generation installation can be substantially lower than those from a fossil-fuel plant. Biofuel emissions can be lowered and efficiency increased by using clean, dry fuels, gasification systems, and other technologies. Many biomass products can also be converted to methane via biodigesters, then burned or further reformed into hydrogen and reacted in fuel cell systems.

We do not propose installing a biomass generator at this time, although we will look at possibilities to use onsite agricultural and forestry products and eventually test alternative biofuels (e.g., algae-sourced biodiesel, food-industry wastes) produced on campus, offsite, or both.

Bulk Energy Storage

As discussed in detail in chapter 3, storing energy is a crucial aspect of energy independence for Greenway College. Bulk electrical storage and recovery lets us decouple the generation of power from its usage—in our case by storing many hours of energy from wind and solar power for use exactly when needed. Pumping water to a higher elevation for energy storage is the most common large-scale stationary energy storage system in the world; in the United States, about 135 such hydro-pumped storage sites are operational with a total power generation capacity of about 18 GW.[13] Although the most economical storage method will depend partly on the landscape where Greenway is ultimately built, for this estimate we will assume that Greenway's site enables the construction of a 2-acre reservoir at sufficient head to

---

13   "Inventory of Electric Utility Power Plants in the United States 2000," DOE EIA 2000 report, accessed July 2007, http://www.eia.doe.gov/cneaf/ electricity/ipp/ipp_sum.html. This data is specifically from Table 17 (http:// www.eia.doe.gov/cneaf/electricity/ipp/html1/t17p01.html).

store 35,000 kWh of electrical energy. This is enough to run the whole campus (assuming an average of 5,000 kWh daily consumption) for seven days—a generous backup supply. We estimate the total cost of such a system at roughly $5 million.

Short-Term Energy Storage

Short-term uninterruptible power supply (UPS) storage provides additional protections for Greenway's power microgrid, reacting to near instantaneous changes in power requirements. Currently, we propose either lithium-ion-based storage or flywheel energy storage. One leading option is for the majority of the UPS backup to come from a long-lifetime flywheel-based system similar to the one from Active Power, Inc. These flywheel systems are commercially available and have superior lifetime environmental performance over battery-based systems, based on both materials (no exotic or hazardous chemicals) and lifetime (thirty-plus years as opposed to closer to five years for most battery-based UPS systems). A 500 kW flywheel system providing about fifteen seconds of coverage (until backup and bulk storage can take over) costs about $75,000.

Backup and Portable Energy Storage

A hydrogen electrolysis, storage, and fuel-dispensing system will be installed at Greenway to provide hydrogen for backup energy storage and portable usage. Hydrogen is sometimes slated as the mobile energy storage medium of the future, because it is low weight and clean burning (when combusted or reacted with oxygen, the only byproduct is water). Hydrogen energy storage has low roundtrip efficiency (20 to 40 percent), but it is a great option for backup generation in place of a diesel generator and for mobile uses. Our hydrogen setup may include an approximately 400 kW power generation unit, providing several hours of backup power. Total cost for such a packaged system (e.g., from Hydrogenics) is about $2 million.

Energy storage for on-campus vehicles will be primarily in the form of on-board batteries with some hydrogen and biofuel vehicles. Electrochemical batteries are currently the best technology for small portable devices and also are used for small and moderate-sized power equipment, including electric vehicles. Proper recycling of battery chemicals and materials is essential to a zero-waste closed-loop system. Currently lead acid batteries claim that "more than 97 percent of all battery lead is recycled",[14] whereas most other batteries have much lower recycle rates. Lithium-based rechargeable batteries—which have a relatively low weight, longer lifetime, and reasonable cost—continue to gain popularity as the technology and costs improve.

Much of the cost of energy storage for transportation will be folded into the cost of buying the vehicular fleet itself. For the initial, core campus, we estimate a fleet of several light trucks, vans, and grounds-keeping vehicles (tractors, mowers, etc.), mostly electrically powered, at a total cost of approximately $1 million.

Altogether, the energy storage systems at Greenway College will provide a highly reliable, totally green stand-alone microgrid and complete working environment, including transportation and maintenance. The systems will be overengineered as they serve not only as a highly reliable working microgrid, but equally importantly as a clean technologies study tool, demonstrator, and campus showpiece. This stand-alone totally green microgrid and complete zero-waste working environment is based on existing technologies but overall will be the first of its kind.

---

14   "Battery Recycling," Battery Council International, accessed October 2012, http://batterycouncil.org/?page=Battery_Recycling.

Wastewater and Solid Waste

The average American directly uses about 100 gallons of water a day, about a fourth of this going to blackwater (toilet water).[15] Wastewater treatment will be accomplished using two natural methods. First, in all buildings, blackwater will be separated from washwater (graywater). Toilet waste will be handled by composting toilets whose end product will be useful rather than burdensome or offensive. The Greenway College composting-toilet (blackwater) system will include about twenty fixtures. The graywater system will use a greenhouse with finishing processing done by constructed wetlands and ultraviolet disinfection.

Anything that one might send to a landfill is "solid waste." At Greenway, solid waste will be eliminated largely through preventive nonuse (e.g., elimination of nonessential packaging), reuse (e.g., of essential packaging, washable utensils), use of truly recyclable products (e.g., pure metals, pure glass), use and composting of biodegradable "clean" products, and producer takeback/deposit agreements. Our vision is to use only products that are designed based on life-cycle analysis and zero-waste policies. Experience shows that products can be designed and manufactured in this manner without sacrificing quality and performance, and even, in many instances, with eventual reduction in cost. Companies that collaborate with us in designing and manufacturing such products will likely benefit in the long term—or the not-so-long term.

Costs for a program to eliminate all solid waste from Greenway College are primarily for operations (including personnel). Startup costs include building space for collection

---

15   "Indoor Water Use in the United States," US Environmental Protection Agency, June 2008, accessed May 25, 2012, http://www.epa.gov/watersense/docs/ws_indoor508.pdf.

and processing of materials to be reused, recycled, or returned to the manufacturer. Much of this space is designed into the initial Plant Operations/Dining Services building, but additional space will be budgeted near the secondary compost facility. Food waste will be composted in an in-vessel system located near the dining hall. A covered building and additional aerated static pile compost facility will be maintained near the greenhouse water-treatment area. These facilities will be landscaped to fit with the natural beauty of the site, will (as mentioned earlier) be designed for strict odor control based on the extensive operating experience of many such facilities around the country, and will provide additional processing and storage space for materials.

The cost of these additional Plant Operations facilities will be on the order of $600,000, depending on the square footage of the site. Overall, solid-waste processing will cost a small fraction of initial construction but will entail an approximate manpower and maintenance cost on the order of $200,000 per year, similar to related activities at a standard institution.

Conservation and Efficiency

The cheapest unit of energy is the one you never use: on a cold day, it is always better to close your windows than to crank up your oil burner. That is why energy experts universally agree that efficient end use is a key part of any rational, cost-effective energy strategy. This is true at every scale, from the household to the nation: dollar for dollar, efficiency investments can reduce cost, emissions, and risk (e.g., exposure to fuel-price volatility) faster than investments in *any* form of energy generation. It is conservatively estimated that the United States as a whole could reduce its energy generation requirements by 23 percent with efficiency and conservation measures that would pay for themselves.[16] Americans presently use about 35 percent more energy to produce each unit of gross domestic product than do Europeans[17]—who intend to increase their already pretty-good energy efficiency by another 20 percent.[18] These numbers matter because higher energy efficiency, far from being a mere amusement for the greener-than-thou, translates to decreased dependence on energy imports, greater resilience against energy price shocks (because of strengthened demand response), lower pollution, lower production costs, and other bottom-line benefits.

At Greenway College, all buildings will feature zero-sacrifice, high-efficiency, long-lifetime systems for lighting, heating, and other technological services that improve efficiency and decrease consumption. Sustainable building practices can affordably reduce energy consumption per square meter (summed over all categories of use) to a small fraction of business-as-usual standards. In general, wise efficiency measures add marginally to start-up costs but more than pay for themselves in reduced operating costs. For reducing electricity consumption, proven zero-sacrifice practices include daylighting and motion-and-light-sensitive, high-efficiency fluorescent and LED lighting; proper sizing of pipes and pumps; thicker in-wall wiring; outside-air ventilation management; proper insulation; passive solar design for heating,

16   "Unlocking Energy Efficiency in the US Economy," McKinsey & Company, 2009, accessed January 9, 2012http://www.mckinsey.com/Client_Service/Electric_Power_and_Natural_Gas/Latest_thinking/Unlocking_energy_efficiency_in_the_US_economy.

17   "World Energy Intensity—Total Primary Energy Consumption per Dollar of Gross Domestic Product Using Purchasing Power Parities, 1980–2006," Energy Information Administration, December 19, 2008, accessed January 10, 2012, http://www.eia.gov/pub/international/iealf/table1p.xls.

18   "Action Plan for Energy Efficiency: Realising the Potential," Commission of the European Communities, October 2006, accessed January 10, 2012, http://ec.europa.eu/energy/action_plan_energy_efficiency/doc/com_2006_0545_en.pdf.

cooling, and hot water; computerized power management; and use of high-efficiency appliances.

Start-up Costs, Operating Budget, and Funding

Let's collect the money figures scattered throughout the earlier parts of this chapter. Real estate will be under $10 million; buildings, approximately $50 million; energy generation—including electricity, space heating and cooling, hot water—around $12 million; energy storage, including cost of an electric on-campus vehicular fleet, around $8 million. Total facilities costs are thus in the ballpark of $80 million. A fundraising goal of $120 million will be set for all design, construction, and facilities costs. Greenway could almost certainly be started for a lower figure, but we strongly feel that the facilities cannot be underengineered without compromising the long-term success of the college. Any funds not used for initial construction will be returned to investors or added to the endowment at the option of the specific donors.

Our proposal includes twenty faculty and thirty-five staff at Greenway College, with an additional five researchers paid half salary and two Greenway Institute staff, for a total expected yearly cost of $4.7 million in personnel. With $2 to $2.5 million for supplies, upkeep, and new and replacement equipment, the annual operating budget checks in at about $7 million.

With approximately 175 students paying $30,000 in tuition, we would only have $5 million in annual revenue. This gives a shortfall of about $2 million—not large or unusual for such an institution. This is what endowments are for. Assuming a 4 percent rate of return, a $50 million endowment would return the $2 million per year; an endowment of $175 million would cover all operational expenses for the same rate of return. Funding will be solicited from private donors and corporate sponsors to purchase the land

and build the first state-of-the-art new buildings with power generation, energy storage, and waste treatment facilities.

In sum, we propose to raise $120 million for initial construction and a minimum of $50 million for an endowment. The total is much less than what it takes to build a large amusement park.

Feast or Famine?

Consider two visions of our future world from the popular and scientific press of today: Famine and Feast. Will we doom future generations to Famine—a world depleted of precious natural resources, with reduced biodiversity, polluted by unnecessarily concentrated toxins and waste products, over-populated and hungry, and with a changed climate due to ever-increasing emissions leading to global warming? Or are we on track for a Feast, in which the world gets better all the time through ever-increasing ingenuity, invention, market forces, and brilliant engineering? One Feast-style forecaster, Julian Simon, predicts that "[t]he material conditions of life will continue to get better for most people, in most countries, most of the time, indefinitely. Within a century or two, all nations and most of humanity will be at or above today's Western living standards."[19]

The Feast forecast is supported by data on health, poverty, hunger, and technology that show long-term improving trends over many generations and even in recent decades. But a Famine fan might counter—recycling an image we introduced in chapter 1—that these data merely document the comfy conditions aboard a car accelerating toward a cliff edge. Are industrial civilization's gains *sustainable?*

Greenway's founders are not qualified to venture a detailed prediction of Earth's planetary future, but we are

---

19   Ed Regis, "The Doomslayer," *Wired*, May 2002, accessed May 24, 2012, http://www.wired.com/wired/archive/5.02/ffsimon_pr.html.

confident that that future lies somewhere between the two extremes of Feast and Famine. We argue that industrial civilization will prove sustainable if we *make* it sustainable. This means putting our research, engineering, business, and policy-making powers toward creating a sustainable Feast for the whole world—a goal both worthy and plausible, though not guaranteed or automatic. We do know that through invention, engineering, and technology, we can design products, processes, and communities that are "green"—zero emission, zero waste, and sustainable—while maintaining or improving performance, using fewer resources, and sustaining human well-being.

Never have conditions been more opportune for the founding of such an institution. Never has the importance of sustainability, energy independence, and efficiency been so widely appreciated by private citizens, corporations, and government. Never have so many tested, cost-effective technologies been available with which to weave a sustainable global future. We believe, therefore, that Greenway College, as a uniquely timely and exciting endeavor, will attract superb faculty and students from the US and the world. Highly qualified professionals in clean technologies will be lured by the prospect of educating enthusiastic students in a hands-on, self-directed, apprentice-style setting, working with local and distant communities, corporations, and other organizations to further exciting technologies. They will be lured by the prospect of creating a healthy, fair, and economic workplace for our talented staff and students. Students will be attracted to our unique, hands-on program and the can-do hopefulness of an institution entirely devoted to enacting, developing, and disseminating the techniques of sustainability. We are therefore confident that the *demand* for a Greenway College is out there from all standpoints: faculty, students, corporations, donors, and communities.

As previously outlined in this book, the technology

already exists to build a totally green college that sacrifices no creature comforts. We thus look forward to moving from the multi-year process of developing this proposal to the joy of designing, constructing, and working at Greenway College. We foresee that Greenway will become one of the world's most respected, talked-about, and emulated institutions.

Contact information for the Greenway founders' team is given at the back of this book. We hope that you will choose to help us move toward the realization of this vision.

Teaching Philosophy and College Governance

I hear, I forget.
I see, I remember.
I do, I understand.

—Chinese proverb

"Tell me and I forget," Ben Franklin wrote. "Teach me and I remember. Involve me and I learn."

Introduction

We now address the educational program of Greenway College: its scope, pedagogy, curriculum, admission criteria, faculty profile, and governance. All these aspects of Greenway's program are designed to provide a challenging, personally rewarding educational setting for both learners and teachers.

A Broad, Narrow Mission

If Greenway tries to be all things to all people, it will fail. That is why its academic program, at least at first, will be limited primarily to engineering, technology, and the sciences, and why in its early years, all students will major in either "sustainable technology" or "sustainable engineering." The distinction between these two will be primarily based on course selection, with the technology degree having a much broader set of options, less mathematical rigor, and different competency exams. Far from being suffocatingly narrow, these fields of study are staggeringly broad already and getting broader all the time.

But why a brand-new four-year undergraduate college? Why not a new department or graduate school in an existing university, a residential community, an industrial company, or some other arrangement?

A brand-new institution is called for, to begin with, because it will be easier to build an institution as pragmatically radical as Greenway without having to negotiate the habits, vested interests, and micropolitics of an existing institution. Second, we look to the four-year college framework because it offers an exciting array of opportunities for improving on the teacher-centered, lecture-based classroom model. This brings us to the question of pedagogy—how to teach.

Student-Centered, Active Learning

Greenway wants its graduates to make things happen. It must therefore confront the tendency of traditional programs to produce graduates who have mastered a lot of information but are too often ill equipped to put it into practice. A famous video shows graduating MIT engineers who cannot figure out how to light an incandescent bulb with a battery and a single wire.[1] Not all engineering graduates are this challenged—but it is clear that Greenway must address not only technology, but teaching methods.

From high schools to graduate and medical schools, lectures are by far the most common method of transferring knowledge from teacher to student. They are, however, a slow,

---

1   "Minds of Our Own," Annenberg Learner, http://www.learner.org/resources/series26.html?pop=yes&vodid=278761&pid=76#. The solution is to touch the electrical contact at the bottom of the bulb's stem to one terminal of the battery, then use the wire to connect the screw thread to the battery's other terminal.

unsure, and often sleep-inducing method of teaching, and are no good at all for inculcating practical skills. The lecture format of a typical college-level science course, supplemented with textbook problems, textbook laboratory experiments, and written exams, is ineffective for many students—one reason for the high attrition rate from such programs. A recent study of reasons for the high attrition from science, math, and engineering majors in colleges found that 90 percent of switchers (students leaving for other majors) and 75 percent of persisters said that the quality of instruction was low, raising issues of pedagogical effectiveness, assessment, and curricular structure.[2] Studies have found that high attrition is encouraged by large class sizes, inaccessible instructors, and uninspiring teaching methods, among other factors.[3] It may be just as effective to give students a textbook and let them read it on their own as to lecture them on its contents.[4] A prime advantage of the lecture method is its cheapness: one can (and many schools do) seat several hundred students at once in front of a single lecturer in, say, chemistry, and let 'er rip. This approach is at the opposite extreme from what is probably the most effective—and expensive—teaching method of all, one-on-one tutoring.

---

2   Peter A. Daempfle, "An Analysis of the High Attrition Rates Among First Year College Science, Math, and Engineering Majors," Educational Resources Information Center, 2002, accessed February 13, 2012, http://www.eric.ed.gov/PDFS/ED465347.pdf.

3   D. W. Knight, L. E. Carlson, and J. F. Sullivan, "Improving Engineering Student Retention Through Hands-On, Team-Based, First-Year Design Projects," American Society for Engineering Education. (paper presented at Proceedings 31st International Conference on Research in Engineering Education, Honolulu, Hawaii, June 22–24, 2007), accessed February 13, 2012, http://itl.colorado.edu/images/uploads/about_us/publications/Papers/ICREEpaperfinalin07octJEE.pdf.

4   Lion F. Gardiner, *Redesigning Higher Education: Producing Dramatic Gains in Student Learning*, Report No. 7 (Washington DC: Graduate School of Education and Human Development, the George Washington University, 1994), accessed February 10, 2012, http://www.eric.ed.gov/PDFS/ED394442.pdf.

A classic image of the traditional method is found at the beginning of Dickens's novel *Hard Times*: "Now, what I want is, Facts. Teach these boys and girls nothing but Facts. Facts alone are wanted in life. Plant nothing else, and root out everything else. You can only form the minds of reasoning animals upon Facts: nothing else will ever be of any service to them. . . . Stick to Facts, sir!"

Since at least the 1950s, researchers have been scratching their heads over how to affordably improve on the obvious drawbacks of the lecture method. New teaching methods that attempt to break out of the lecture-based, teacher-centered scheme have in many cases (though not all) been successful. Yet these new methods remain rare, partly because their proponents are trying to introduce them *within* institutions committed for decades or centuries to the old lecture-based format. In the face of resistance from entrenched lecture-centered majorities, change is inevitably slow and difficult. Greenway sidesteps this problem: it will implement a flexible, nondogmatic range of the best evidence-based educational methods across the board and from day one.

The gist of our approach can be summed up in the phrase "student-centered, active learning." Despite negative stereotypes that contrast tough, old-fashioned methods with muddle-headed, permissive fads, student-centered, active learning does not mean lowering standards, coddling slackers, or inflating grades. It is completely wrongheaded—and the scientific evidence bears it out—to think that the suffering student learns more than the engaged self-motivated active learner. According to a 1999 National Research Council report,[5] a person's ability to recall "a rich body of knowledge in a subject

---

5   John D. Bransford, Ann L. Brown, and Rodney R. Cocking, eds,, *How People Learn: Brain, Mind, Experience, and School*, Committee on Developments in the Science of Learning, Commission on Behavioral and Social Sciences and Education, National Research Council (Washington DC: National Academy Press, 1999).

matter" is key to problem-solving ability in mathematics and the sciences. By themselves, though, crammed-in facts are practically useless. As educators, most of what we accomplish by such cramming is to weed out those students who cannot or will not absorb large masses of undigested fact. When we graduate the remainder, we congratulate ourselves on having taught a rigorous program, and the students are happy because they have their piece of paper that certifies their intelligence and work ethic. But can they light the light bulb?

According to the National Research Council, factual subject matter must be closely tied to an understanding of how that subject matter is interconnected, and how it is may be applied to solve new or difficult problems. The answer: *student-centered, active learning.*

Student-centered learning is designed around what students do, rather than around what teachers do. "Active learning" is a term of educational art for a class of student-centered teaching approaches. According to one review of the literature, active learning is "any instructional method that engages students in the learning process. In short, active learning requires students to do meaningful learning activities [often in the classroom] and think about what they are doing."[6] Active-learning approaches can include lecturing, especially upon student request, but must break it up frequently (at minimum every ten to fifteen minutes) with activities and problem solving. They feature collaborative and cooperative learning, where students work in small groups toward a common goal, and problem-based learning, where problems are introduced *before* solution techniques to motivate self-directed study. For example, among other activities, students are assigned difficult problems that they

---

6   M. Prince, "Does Active Learning Work? A Review of the Research, " *Journal of Engineering Education*, 93(3) (2004): 223–231, accessed February 13, 2012, http://www4.ncsu.edu/unity/lockers/users/f/felder/public/Papers/Prince_AL.pdf.

must learn to solve on their own. They can consult textbooks or the Internet, perform experiments, or ask the professor to provide lectures, information, or clarification—but they must come up with a solution on their own.

A large body of research shows that active learning can be a radical improvement over traditional instruction (figure 6.1).[7] Many of these data come from introductory classes at large institutions, classes which tend to be large. At a small college with small class sizes, additional active-learning techniques that can further enhance learning and motivation become feasible. For example, in a smaller class, it is easier to notice when students are not learning, not engaged, or not motivated, and make midcourse corrections.

Average College and University Results

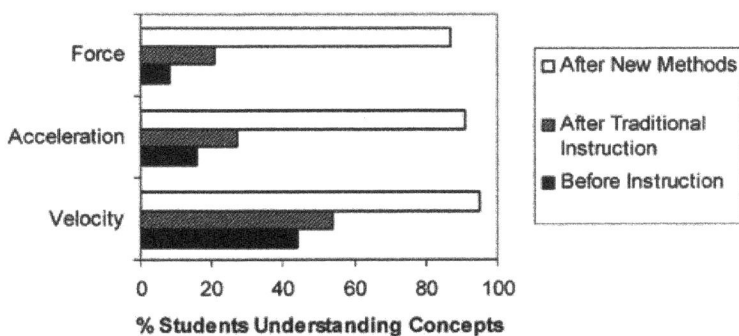

Fig. 6.1. Active-learning versus traditional instruction for improving students' understanding of three basic physics concepts (force, acceleration, and velocity).[8] The new methods produced very high levels of understanding for all concepts, regardless of understanding levels prior to instruction.

7   See the large annotated list assembled by J. E. Froyd, a professor at Texas A&M, at http://www.wmich.edu/science/facilitating-change/Products/FroydPoster.pdf (accessed February 13, 2012).
8   P. Laws, D. Sokoloff, and R. Thornton, "Promoting Active Learning Using the Results of Physics Education Research," *UniServe Science News* 13 (1999).

Finally, in a small-college setting, a focus on project-based learning and apprenticeship models offers additional forms of active learning. My personal experience in teaching project-based learning courses—building solar-powered boats and cars—for first-year engineering students validates the research: first-year students achieved remarkable results. Based on this experience, I feel strongly that many engineering topics could be taught in a project-based manner and that for many self-motivated, hands-on learners, this method is optimal. And, as active-learning proponents often point out, if one can't think of an application or hands-on project to teach an engineering concept, that concept will probably not be relevant to most future engineers anyway.

Involving students in engineering has a long history. In the 1930s, for example, mechanical engineering education at the University of Minnesota was highly hands-on. Students designed and built internal combustion engines from foundry to final assembly and testing. In keeping with the insights and principles of student-centered, active learning, Greenway's curricular design will be centered on a peer- and faculty-mentored, hands-on, project-based, problem/solution-centered approach to mentoring students within the frame of their individual education plans. Each student will be involved, starting with their first semester, in projects analyzing current methods in energy and other sustainability-related technology areas. (Research shows that hands-on design projects increase student retention.[9]) In later years of their program, students will be challenged to propose improvements on existing technologies or to perform depth analysis of the performance of current technologies, either on-site or working with industrial or municipal partners.

Students will all take required yearlong courses in "Engineering and Sustainable Technology" and will have the oppor-

---

9   Knight et al., op. cit.

tunity to take course modules similar to those in traditional programs—we describe the curriculum more fully below—but much of their learning will be independent or group-directed study modules. Lecture will be kept to a minimum; discussion-type learning will dominate in the classroom; students will be responsible (with help from faculty and peer mentors) for choosing many of their own projects and academic specializations. Students will continually modify their study plan with their mentor group and maintain a portfolio of their studies and accomplishments. Additionally, students will be strongly encouraged to seek apprenticeship arrangements in specific applied topics.

As Greenway College evolves, new sustainable facilities may be added or existing buildings modified to make them totally green—resource-neutral or even resource-positive (centers of energy or other production). These on-site events, though only a part of the college mission, will provide opportunities for new technologies to be explored by researchers and students. Collaborations within industry and public works will be extensive. Faculty research will be structured to bring teachers together with students, rather than draining teachers' hours from teaching.

Curriculum

Entering students will be provided copies of textbooks in core engineering disciplines. Students will take a common Engineering and Sustainable Technology course sequence through all four years to guarantee a shared core of competence, and will fill their remaining schedule with course modules that include standard faculty offerings; special-interest topics in either independent or directed-group study; project-based courses; and apprenticeship modules. Each student will have an advising committee that will assist them in laying out an individualized program of study. Students will also maintain work portfolios in concert with their plan of study. Students seeking an engineering degree will take a set of required general-competency exams during their first three years (carefully timed and structured to minimize exam stress). Final decisions on the academic program will be determined by the founding faculty.

Below, we discuss each of these components in a little more detail.

Textbooks. In most teacher-centered, lecture-based settings, course content is compartmentalized by topic. Thus, students need only one textbook at a time, and learn material in a set order and hierarchy. Where students set their own educational path and undertake projects that call on material from many disciplines throughout their education, they will need access to many books at once. Students have complete access to the solution archives from their first day on campus, encouraging self-directed study.

Engineering and Sustainable Technology Course Sequence. All entering students will take a yearlong Introduction to Engineering and Sustainable Technology course that will cover topics important to all branches of modern engineering, such as common engineering solution techniques, software, instrumentation, mathematics, statistics, design, optimization, project management, problem solving, teamwork, ethics, history, current status, disciplines, and technical presentation/writing skills. Topics in green technologies, lifecycle analysis, design for the environment, and future challenges for sustainable technology will be addressed. Students will undertake several projects that connect these topics to practice. This course will provide a strong, consistent background for all students, introducing them to the major concepts of engineering and green technology while uniting the students as classmates.

Engineering and Sustainable Technology courses continue in every year. In addition to engineering and green technology topics and projects, the upper-level courses will cover logistical material related to student plans, portfolio, curriculum, competency testing, and capstone projects.

Course Modules. All courses except Engineering and Sustainable Technology will be offered as multiples of three-week modules. A catalogue of standard modules will be available, but new modules may be introduced by students or faculty at any time when interest is garnered. Students will be expected to regularly meet with their advisory committee to set and evaluate their plan toward graduation. Students are expected to have a full schedule at all times, but are encouraged to sign up for independent study and off-site modules. Proposed modules include the following:

- *Traditional engineering topics.* Motion, Newton's Laws, Energy-Method Solutions, Geometrical Optics, Derivatives, Ordinary Differential Equations, Electric Fields, Basic DC Circuits, etc.

- *Special-interest technical topics.* Wind Turbine Science, Solar Photovoltaics Science, Flywheel Technology, Traditional Waste Management Systems, Wastewater Technology, etc. Some of these special topics will be set up as *project-based courses,* centered on building a device or system in the context of in-depth technical study.

- *Apprenticeship module.* Students can opt to receive credit for working closely with faculty, staff, or local (or remote) engineers or technologists to learn current practices, a technology, or an area of research. The student must propose learning objectives, format, and assessment criteria to an advising professor.

- *How Things Work sequence.* These qualitative introductory courses, including Motion and Mechanics, Thermodynamics, Electricity and Magnetism, Light, and others, will cover the qualitative content and applications of a topic through description, animation, demonstrations, reverse engineering, and small projects. More rigorous courses in each subject area will follow.

- *Nontechnical modules.* Students will be encouraged to take nontechnical modules based on their interests—history, music, art, languages, botany, literature, philosophy, psychology, religion, or others. These modules may be taught by Greenway College professors or local community residents with special interests, taken as self-directed study individually or in a group, or taken at neighboring colleges.

Plan of Study and Advisory Committee. A major part of the Greenway College experience will be each student's development of a *plan of study*, assembly of a *portfolio of work*, and completion of a *capstone project*. On arrival, first-year students will be paired with a teacher and upper-class mentor. Midway through their first year, the student will begin to write their four-year plan, and by the end of the year will have produced a draft plan and have begun to approach individuals to serve on their advisory committee.

The plan will specify between half and three-quarters of the student's courses in advance, providing structure while leaving space to explore new topics. Advisory committees will consist of three or four members, one being the primary advisor. Each student's committee will form midway through their second year and will include up to two faculty members, up to two outside professionals, and up to two peers.

At the end of their sophomore year, each student will finalize (with their committee's approval) a plan of study for their remaining time. Students will have broad freedom to shape their own study plans, but in all cases these plans must be approved by the students' advising professors.

Portfolios. The portfolio will be a collection of representative coursework, project reports, evaluations, presentations, software, and materials that summarize and demonstrate the student's knowledge, talents, projects, path of study, and work experience.

Capstone Project. In their senior year, each student will complete a capstone project. This will be similar to a senior or master's thesis, with course-module load adjusted to the depth of the investigation or project. Evaluation requirements for the project will be determined by the student in cooperation with their advising committee. The only universal requirements are midway and final progress reports and presentations.

Admissions

Greenway's success will depend not only on effective teaching, but on attracting and selecting the right students. Bright, hardworking students will be attracted to Greenway College by its unique mission and ambitious learning structure. But we must try not to admit students that would be better served by a traditional university setting—those who would thrive in lecture-based courses and whose style is not hands-on. (Greenway might not be the best place for a budding genius of pure mathematics.) Admissions policies may include one year of nonacademic, preferably post-high-school, experience—work, military service, volunteering, community theater, the whole range of meaningful nonacademic pursuits. We think that it is important for students to develop some experience outside of the academic bubble. This requirement will encourage selection of students capable of taking control of their own education.

We may also set our sights on decentralization of the admissions process through donor organization selection criteria. Here's what we mean by that: when endowment fundraising goals have been met, all students entering the college will have direct sponsorship from a donor organization (or group of donors). The people or companies providing the donations have the option to provide a list of selection criteria and/or to provide a list of five to twenty qualified candidates. These must not discriminate on the basis of race, gender, religion, ethnicity, or political belief. Members of the college will make the final acceptance decisions.

The intent is to have companies and individual donors around the country (and perhaps internationally) helping us to select a diverse, unique set of incoming students without the need for a large, costly admissions department. This arrangement will give donors strong say in the makeup of the college while still keeping control over incoming class makeup.

Finally, we are aware that top-notch college applicants increasingly expect an international aspect to their college experience. Greenway will offer students the option of spending a half- to one full school year at another college or university, in traditional study abroad, or in nontraditional combined work study/volunteer programs. Greenway College will admit half- or full-year visiting students from other colleges, universities, and possibly industry on a competitive basis. Visiting students may make up at most 15 percent of the student body.

Faculty

The job of a good teacher is to help the student to discover their own interests, strengths, and abilities, channeling the power of curiosity to maximize learning and fulfillment. A good school encourages teachers to pursue their passions and share their excitement with the students—not corner a teacher into teaching boring material to bored students.

The first wave of faculty and staff will be hired one to two years prior to the arrival of the first student class, and will design the day-to-day details of the college before the college opens its doors to students. Attendance by the first student class will coincide with near or full completion of the first campus buildings. We aim to open Greenway within three years of reaching the $80 million fundraising level; this time-table is short enough that, while the full contingent of faculty (about twenty full-timers) and staff will be on campus by the fourth student class, the first operating year may see as few as five to ten faculty and a first class of twenty to thirty students.

One sure-fire way to attract top faculty is to pay well. At Greenway College, pay for faculty members will be good—we project an average of $130,000 including insurance and other expenses—but not overwhelmingly so. Pay will also be relatively flat, with relatively small increases for experience and accomplishments. By not offering large monetary incentives, we will lose some excellent candidates, but we expect ample motivation to come from other sources besides money—such as the opportunity to have an extraordinarily effective teaching and research career with a discernible impact on the reengineering of the world. By making our pay system transparent—all pay scales (with names redacted) and financial policies will be available to all community members—we will discourage people from applying who see that Greenway does not meet their current or future needs.

Faculty and staff will be hired on a series of short-term contracts that sequentially increase in duration; we will not offer traditional tenure. The tenure system protects valuable faculty at institutions where powerful people can fire those who do not conform to their ideals; at Greenway, we will strive to create a democratic system in which no one person or small group of individuals has excessive power, to prevent such abuses and thus the need for their traditional remedies. We will retain, however, the ability to lay off persons who act to the detriment of the college. The proposed governance of the college will be discussed in more detail later in this chapter.

We add a word here about accreditation: recognition, by some recognized body, that a college or university meets certain standards. Accreditation by organizations recognized by the US Department of Education is the public's ultimate guarantee that a school is not a fraud or scam. All serious colleges are accredited, and Greenway shall be too. Certain alternative colleges, such as Hampshire and Marlboro, have already been fully accredited by the New England Association of Schools and Colleges; both are highly progressive, with students designing their own programs. We are therefore sure that Greenway College will be able to promptly meet the requirements for general accreditation without sacrificing its educational freedom and mission. Engineering accreditation through the Accreditation Board of Engineering and Technology (ABET) may not take place until the educational program is well in place and tested, likely several years after being established. Accreditation by ABET is not overly restrictive: the group modernized its requirements in 2000, and now encourages new engineering education programs and schemes. In many ways, the Greenway College curriculum is exactly what educational researchers in ABET would love.

Governance and Administration

Last, and hopefully not least, we come to the question of how the college will run itself. As with pedagogy, we propose to break with traditional methods in some respects, but reasonably, not rashly.

Many private colleges and universities today are set up in quasi-democratic arrangements, with a faculty senate and student senate. However, these entities typically have limited power to run the institution: real power rests with the board of trustees and the president. This setup actually works well in most cases, but it is only superficially democratic.

Our intent at Greenway College is to create a fair balance

of real, operational power among donors, faculty, staff, students, and local community members. Much care will be taken to keep the college's self-governance democratic, maintaining operational flexibility, fairness, and the ability to act quickly and decisively, and preventing "poisonous" elements (e.g., people who develop personal obsessions or grudges) from blocking up the system. Major donors to Greenway College will, we believe, appreciate the college's transparent administrative structure and democratic commitment, which echoes basic commitments of American governance. Indeed, the US constitution is a most effective plan for fair democracy, and on it we base our administrative proposal. A college is not a country—it is smaller, and different in nature—but much of the wisdom of the Founders can be transposed into this setting.

At Greenway, therefore, an Executive Council corresponds to the presidency in the US government—the top authority of the executive branch. The legislative branch consists of a Faculty Senate and a House of Representatives, the latter having members from the faculty, students, staff, and community. A Judicial Board will consist, in the college's early years, of five founders of the college, likely comprised of donors, designers, or early faculty. This branch is charged with maintaining and expanding on the original vision, goals, and mission of the college.

The day-to-day business of the college will be executed by a somewhat traditional administrative structure. Additional to the staff and faculty already listed elsewhere, administrative staff includes a provost, administrative assistant to the Executive Council, and half-time treasurer.

Executive Council. The Executive Council approves (or vetoes) measures passed by the Senate and House, proposes changes and courses of action, and generally leads the college. Also, it is the primary fundraiser and typically makes recommendations for hiring, expansions, and future plans. It can take certain actions without prior approval of the Senate and House, but all actions will eventually require approval.

Faculty Senate. The Faculty Senate and House of Representatives have control over all legislative issues, including curriculum, community outreach, projects, and some finance. Bills must pass both the Senate and House and be approved by the Executive Council (with provisions for veto overturn). Decisions of the legislature are subject to review by the Judicial Board.

All faculty members are members of the Faculty Senate, with one vote each. The dean of faculty assumes a role similar to the vice president in the US Senate: presiding, but voting only to break ties. All voting rules, such as two-thirds override of a veto, impeachment, and association with the House, parallel those of the US Senate where possible.

House of Representatives. The House will consist of representatives from staff, faculty, the student body, and the local community, and will otherwise mirror the US House of Representatives. At the outset, we suggest four representatives from each group for a total of sixteen members. (As the college population grows, the House may grow proportionately, as does the US House.)

Judicial Board. Founding members of the college will populate a five-person Judicial Board. These members will elect a new member whenever a member wishes to step down; members can also be removed by impeachment by the Congress, in which case a new judge will be selected by the Executive Council. The Board's job is to make sure the college sticks to its mission and course. It can strike down measures that it sees as outside of this mission. In this manner, it acts as a watchdog for the long-term interests of the college. It is charged with keeping the college government lean and responsive, preventing overlegislation and promoting academic freedom.

We believe that a democratic, checks-and-balances form

of college government modeled on the world's oldest functioning, written constitution, that of the United States, will benefit our educational goals. Students and faculty will know themselves as stakeholders of Greenway College, academic citizens rather than mere customers or employees. We believe that as such they will act more responsibly while members of the Greenway community—and donate more generously as alumni of it!

Concluding Thoughts

Some students always manage to learn regardless of teaching style and curriculum, and some teachers always manage to teach rigorous, enjoyable courses even in the most straitjacketed, lecture-based settings. But this doesn't mean that pedagogy doesn't matter. Many students—some of them apparent success stories—are embittered and discouraged by tedious, abstraction-stuffed, needlessly harsh science and engineering programs. It is not even unheard-of for programs to announce to first-year students that it is *policy* to drive a third of them to drop out in despair by the end of their sophomore year. And even the most entertaining professors fail to reach large numbers of students who go on to become fine technical professionals. The cumulative human and monetary cost of generations of such waste is staggering!

Therefore, we must not settle for old methods because they do not fail completely and in every case: that is not good enough. We must provide a learning framework that minimizes useless stress while promoting rigorous excellence. At Greenway we will do our best to help every student we admit to meet our high standards for graduation; we will strive to build an educational program that encourages students to perform at their best while finding their studies fulfilling, enriching, enjoyable.

At the same time, Greenway will attract the finest teachers and support them in integrating teaching with research. Nothing is more satisfying to a good teacher than to see students become engaged, self-motivated, and interested in the material. Thus, we strongly believe that Greenway's proposed program will bring out the best in both students and faculty.

Greenway will not be a degree mill. It will be an accredited four-year college, yes, but also a center of technological excellence, a thought leader in an explosively growing field, a catalyst for change, and a vibrant, democratic community.

Through applied know-how, a better world is possible.

The Other Green

The lack of money is the root of all evil.

—Mark Twain[1]

So far, this book has set forth a vision for Greenway College that we feel is practical, timely, and uniquely important to this country's technological and educational future. Greenway College, when brought to life, will take a leading role in green technologies, sustainable engineering, zero waste, and the modernization of scientific and engineering educational methods.

But a vision only becomes reality when it is shared by a critical mass of people: we cannot make Greenway real without the support of many donors, large and small. We hope, of course, that you, dear reader, will be one of those donors—hence this book! Imagine how it will feel, someday, to see Greenway come to life: to walk the sidewalks of a finished, working campus, some brisk September day a few years hence, talking with the first cohort of students, faculty, and staff, all sparking with eagerness to help engineer our sustainable future.

Before giving, you would probably like to know how your money will be used to make Greenway real. This chapter,

---

1   Mark Twain, *More Maxims of Mark*, ed. Merle Johnson (New York: privately printed, 1927), accessed February 29, 2012, ftp://arcade.demon.co.uk/data/085000/087050/087096.

therefore, outlines the fiscal foundations of the college. It is not a formal prospectus, but is our present best projection of what it will take to get Greenway on its feet and keep it there. The ultimate details may vary slightly, but not the overall scope and shape of what we propose. Greenway must, and shall, engineer its own financial security as rigorously as its students and teachers would engineer anything else.

The Basic Goals

Our fundraising goal to start Greenway College is **$170 million**, including $120 million for initial construction and $50 million for an initial endowment. This initial budget is less than of a medium-large theme park: Circus World, Florida, and Six Flags Great Adventure, New Jersey, each cost $210 million (in 2007 dollars).[2] Endowment fundraising of an additional $125 million—total fundraising goal of **$295 million**—would fully fund yearly operation of the college.

Defined phases of work will begin as soon as fundraising subthresholds are reached:

- Prior to the $1 million level, initial open donations will fund design and fundraising.

- At $1 million, one to three detailed competitive design plans (complete with proposed location) will be developed and posted for public viewing and response on the Greenway College website.

- At $10 million, the site-purchase process will commence.

---

2   Kelly T. Kaak, "Theme Park Development Costs: Initial Investment Cost Per First Year Attendee—Historic Benchmarking Study," Rosen College of Hospitality Management University of Central Florida (paper presented at 16th Graduate Students Research Conference, 2011, Houston, Texas, 2011), accessed February 29, 2012, http://scholarworks.umass.edu/gradconf_hospitality/2011/Presentation/100/.

- At $80 million in validated initial construction funding, designs and bids will be finalized and construction of the college will commence.

As a fundraising challenge and incentive, we propose that Greenway College will be *fully operational and will welcome its first students* within three years of achieving $170 million in total donor funding (including a minimum of $100 million in construction funds).

Ways to Give

Some donors will be glad to fund start-up costs, but others may wish to donate only if and when the financial security of Greenway College has been established. Some may wish their money to go straight to endowment rather than construction, and some the reverse. Since the donor is always right, we propose four donor options, three for construction (including real estate, legal, and other costs) and one for endowment.

1. *Open donation (construction).* Available immediately to the Board of Greenway College for any purpose related to the design, funding, and founding of the college.

2. *Escrow donation (construction).* Deposited to a neutral bank in an interest-bearing money market account. No funds to be drawn until $80 million in validated initial construction funds have been received. At any time before official start of college construction, these funds (plus interest) can be withdrawn by the donor with no penalty.

3. *Targeted donation (construction).* Donations earmarked for specific purposes named by donors (e.g., construction of a memorial room or structure), also deposited in a neutral bank until use. Donors may be

involved in naming or design of structures targeted by major donations and can withdraw funds at any time prior to the start of construction.

4. *Endowment donation.* The endowment will provide funding for college operations based solely on compounded interest. Donations are deposited in a neutral bank until the first class of students is being recruited and can be withdrawn at any time prior to recruiting of the first class.

Incentives!

Like many colleges and universities, Greenway will offer acknowledgement incentives to donors based on giving level.

A Founders' Garden will be situated near the center of the campus (see figure 7.1 for one possible realization). It will include a Founders' Wall displaying donation plaques, with a message (e.g., name) from all donors above $100 and a founding declaration, which reads in draft as follows:

We the founders of Greenway College hereby establish this institution to promote sustainability and environmental responsibility, while developing and forwarding technologies that support such ends and, at the same time, provide shelter, work, and an education for our students, faculty, and staff; this place should be a home for these people, our extended family, that is both comfortable and just, and furthers for these people the best opportunities for obtaining knowledge, liberty, and happiness in a safe and equitable environment. We endeavor to inspire those who pass through this garden and read this message to strive to create a better world that is sustainable; clean; filled with freedom, liberty, and happiness for all; and better in all ways than when you first passed through it.

Fig. 7.1. Artist's rendition of preliminary concept for Founders' Garden and Founders' Wall.

As is common in college fundraising, large donors will have the option to name certain rooms, buildings, facilities, and natural areas. Depending on donor generosity and interests, naming may apply to a commemorative plaque, laboratory or classroom, or an entire building or park. Some donors may wish to fund a wind turbine, a solar array, or other piece of crucial equipment (first come, first served!).

Endowment Fractions

Compound interest from an endowment of $175 million would support all operating costs and salaries for twenty-five faculty, thirty-five staff members, and up to 175 students. With lower endowments, student tuition will fund operations. Donors may give any amount to the endowment, small or large, but in one scenario we propose to pool (or divide) donations into 175 increments of $1 million—one per student. Each student would be associated with a specific donor increment and may act in an apprentice-type relationship with that donor or donor group. Donors may include companies, organizations, and individuals that can both act as a mentor to the student and benefit from the student's experience at Greenway College. In these cases, students will report monthly to donors, may work on collaborative projects, and may work at summer internships for the donor company.

The $1 million increments can be made by one company or individual, or can consist of a group of donors, either known to one another or completely unknown. All endowment donor groups will be acknowledged in the opening hall of the main building, and founding endowment will be acknowledged in the Founder's Garden with the construction donors. The link of the donor to each $1 million endowment fraction is good for twenty years.

Endowment donors or donor groups may take advantage of the following three opportunities: admissions input, update reports from students, and student hiring.

1. Donors may choose to contribute to admissions criteria and to narrowing the applicant pool for a given endowment fraction. This decentralization should make the admission process more personal and diverse (although donors' admission criteria must not discriminate based on race, religion, or the like). A donor who wishes to participate in admissions must select from the applicant pool five to twenty students of roughly comparable ability. From that list, Greenway College will make a final selection. Prospective students must meet general Greenway admission requirements, including one year of experience beyond high school, and a minimum level of achievement in high school and on the SATs, but donors will have significant input into determining the qualifications of the student they are supporting. The college will reserve—though, we hope, rarely exercise—the right to reject any candidate, in order to guard against nepotism and discrimination.

2. Donors may choose to receive monthly and yearly progress updates from the student linked to each endowment fraction. Monthly one-page reports will be a required part of the student portfolio, serving both as an update for the donor group and a progress report by and for the student.

3. Donors may employ students in the summer, after their first, second, and third years. Internships must be related to science and engineering and comply with terms of fairness specified by Greenway College.

Our admissions and endowment model is set up to encourage a sense of responsibility in both student and donor. Students should assume that it is their responsibility to their donor group to perform well at the college, as donors are essentially providing $200,000 to the student's education (not including tuition, meals, and housing) over the four years at Greenway. Second, in cases where donors have a specific emphasis, students can find a sense of purpose and a unifying direction to their education. In many instances, students will be apprenticing with the donor group—learning skills and studying subject matter that the donor feels important. In some instances, the student may eventually be hired upon graduation to fill a related role within a donor company.

Upon graduation, the student has no fixed requirement to work for the donor company if offered employment. As a matter of stated policy, however, Greenway College literature to the student will suggest that students should strongly consider two years of employment at their donor organization if offered a comparable position to other employment offers garnered by the student.

This system will include provisions for cases where students wish to buy out of working for their donor, or where donors are dissatisfied with student performance.

Please Inquire

We have tried to keep this financial chapter as brief as possible, but many more details must be settled before an enterprise as complex as Greenway College can be funded and launched. Readers who want more details about how money will be handled, tracked, and spent are urged to contact the author for direct answers to their questions.

Conclusion

The need for a sustainable economy has never been clearer, nor public support for such technologies broader. Despite the ups and downs inevitable in any innovating market sector, renewable energy is now often the least expensive form of new generation. Energy-storage technologies are swiftly approaching their own day in the sun (so to speak), with inevitably transformative effects on grids and energy supplies. Zero-waste methods are proving their feasibility and worth in case after case; they are, we now know, both practicable and economic. Tens of thousands of net zero-energy buildings, from single-family dwellings to the massive offices and laboratories of the National Renewable Energy Laboratory in Colorado, have proved that net zero-energy buildings can be workable and affordable.

Throughout the industrialized world, the ways we produce, use, and reuse materials and energy are changing. They are changing because technologically, it is *time* for them to change; because it is cheaper for them to change than to stagnate; and because even the proudest, most successful civilization that cannot or will not *learn* is too much like a luxury car speeding toward the edge of a cliff.

Yet we have only begun. Our ingenuity has barely scratched the manifold challenges of sustainability. Implementations and knowledge centers are still scattered, regulations and incentives are too often still perverse, and many technologies are still imperfect, clunky, expensive, or need-lessly complex. Greenway can help change all that, and you can help Greenway. There are lifetimes of work to do, and the sooner we begin them, the better.

The time is right for Greenway College.

Author Contact Information:

Troy McBride
mcbride@greenwaycollege.org

About the Author

Troy McBride switched his engineering career path from medical imaging to sustainable energy after conceiving Greenway College in 2000. Since then he has worked to gain experience relevant to starting Greenway College—including working as a professor of physics and engineering for eight years and starting two renewable energy companies. Troy received his doctorate at Dartmouth College and was a tenured associate professor of physics and engineering at Elizabethtown College in Pennsylvania, where he taught a dozen different technology and science courses, including several on sustainability, and helped lead a successful drive for departmental accreditation by ABET. He has since cofounded and served as chief technology officer for two clean-technology companies, Norwich Technologies and SustainX. Norwich Technologies, as of 2020, employs twenty-five people and has annual revenues of over $20 million. SustainX raised more than $40 million in funding from partners including DOE, NSF, Polaris Venture Partners, and General Electric. Troy has direct experience in start-up organizations, management, sustainable energy technology design, project development, and engineering education. He is pleased to share his vision for Greenway College and looks forward to making Greenway a reality with your help.